PUT

AF692269

RAPRA REVIEW REPORTS

A Rapra Review Report comprises three sections, as follows:

1. A commissioned **expert review**, discussing a key topic of current interest, and referring to the References and Abstracts section. Reference numbers in brackets refer to item numbers from the References and Abstracts section. Where it has been necessary for completeness to cite sources outside the scope of the Polymer Library database, these are listed at the end of the review, and cited in the text as a.1, a.2, etc.

2. A comprehensive **References and Abstracts** section, resulting from a search of the Polymer Library database. The format of the abstracts is outlined in the sample record below.

3. An **index** to the References and Abstracts section, derived from the indexing terms which are added to the abstracts records on the database to aid retrieval.

Source of original article → *Item 1*
Macromolecules
33, No.6, 21st March 2000, p.2171-83

Title → **EFFECT OF THERMAL HISTORY ON THE RHEOLOGICAL BEHAVIOR OF THERMOPLASTIC POLYURETHANES**

Pil Joong Yoon; Chang Dae Han ← Authors and affiliation
Akron,University

The effect of thermal history on the rheological behaviour of ester- and ether-based commercial thermoplastic PUs (Estane 5701, 5707 and 5714 from B.F.Goodrich) was investigated. It was found that the injection moulding temp. used for specimen preparation had a marked effect on the variations of dynamic storage and loss moduli of specimens with time observed during isothermal annealing. Analysis of FTIR spectra indicated that variations in hydrogen bonding with time during isothermal annealing very much resembled variations of dynamic storage modulus with time during isothermal annealing. Isochronal dynamic temp. sweep experiments indicated that the thermoplastic PUs exhibited a hysteresis effect in the heating and cooling processes. It was concluded that the microphase separation transition or order-disorder transition in thermoplastic PUs could not be determined from the isochronal dynamic temp. sweep experiment. The plots of log dynamic storage modulus versus log loss modulus varied with temp. over the entire range of temps. (110-190C) investigated. 57 refs. ← Abstract

GOODRICH B.F. ← Companies or organisations mentioned

Location → USA

Accession no.771897

DOCUMENT DELIVERY SERVICE

Almost all of the documents which are listed in the *References and Abstracts* section are available in full text form, as photocopies or pdf files from Rapra Technology Ltd's Document Delivery Service. Documents can be delivered by a variety of methods, including email, post or fax. Customers may pay for individual copies at the time of ordering by credit card or alternatively open up a deposit account. See the back of this report for further information.

Please contact the Document Delivery Department for availability, current prices and delivery methods.

Document Delivery Department
Rapra Technology Limited, Shawbury, Shrewsbury, Shropshire SY4 4NR, United Kingdom
Telephone: +44 (0)1939 250383 Fax: +44 (0)1939 251118 Email: documents@rapra.net

RAPRA REVIEW REPORTS VOLUME 13

Series Editor Dr. S. Humphreys, Rapra Technology Limited

Rapra Review Reports comprise a unique source of polymer-related information with useful overviews accompanied by abstracts from hundreds of relevant documents. A Rapra Review Report is an excellent starting point to improve subject knowledge in key areas. Subscribers to this series build up a bank of information over each year, forming a small library at a very reasonable price. This series would be an asset to corporate libraries, academic institutions and research associations with an interest in polymer science.

Twelve reports are published in each volume and these can be purchased individually or on a subscription basis.
Format: Soft-backed, 297 x 210 mm, ISSN: 0889-3144

Order individual published Rapra Review Reports (see the following pages for a list of available titles), or purchase a subscription to Volume 13 (12 issues).

ORDER FORM

Title of Publication	Price £/$/€
I would like to order the following Rapra Review Report(s) at £80 / US$120 / €136 each Report Number(s) .... *(please state quantity if more than one)*	
Please add postage at the following rates: UK £5 total, Overseas £7 / US$11 / €10 per item	
Subtotal:	
I would like to order .... subscription(s) to Volume 13 of the Rapra Review Report Series at £630 / US$882 / €1071 each	
Please add postage at the following rates: UK £35 total, Overseas £65 / US$110 / €110 per subscription	
Total Order Value:	

All prices are subject to change and orders will be charged at the price indicated on **www.polymer-books.com** on the date of processing

- ☐ Remittance enclosed
- ☐ Please invoice my company
- ☐ Please charge my credit card

(Please make cheques payable to Rapra Technology Ltd. in £ Sterling/US$/Euros via UK banks only or make payment direct to Account No: 05625025, Sortcode: 55-50-05, National Westminster Bank Plc, 8 Mardol Head, Shrewsbury, Shropshire, SY1 1HE, UK)

American Express / Visa / Mastercard *(delete as appropriate)*

Card Number:

IMPORTANT - Value Added Tax (VAT)

The above prices do not include VAT. Customers in EU member countries may be liable to pay VAT if their Registration Number is not supplied. Please enter your EU Registration Number (VAT - BTW - IVA - TVA - MWST - MOMS - FPA) below:

VAT Number:

Signature: ____________ Exp. date: ____________

Issuing Bank: ____________

Cardholders Name (as on card): ____________

Cardholders Address: ____________

Postcode: ____________ Country: ____________

Telephone: ____________ Fax: ____________

Company PO#: ____________

Full Name: ____________

Company: ____________

Job Function: ____________

Delivery Address (if different from Cardholder's Address): ____________

Postcode: ____________ Country: ____________

Telephone: ____________ Fax: ____________

If you would like to receive regular electronic updates informing you of new titles and offers please enter your E-mail address below.

E-mail: ____________

Please Return to: Publications Sales, Rapra Technology Limited, Shawbury, Shrewsbury, Shropshire SY4 4NR, United Kingdom

Tel. +44 (0)1939 250383
Fax: +44 (0)1939 251118
E-mail: publications@rapra.net

www.rapra.net

Previous Titles Still Available

Volume 1

Report 3 **Advanced Composites**, D.K. Thomas, RAE, Farnborough.

Report 4 **Liquid Crystal Polymers**, M.K. Cox, ICI, Wilton.

Report 5 **CAD/CAM in the Polymer Industry**, N.W. Sandland and M.J. Sebborn, Cambridge Applied Technology.

Report 8 **Engineering Thermoplastics**, I.T. Barrie, Consultant.

Report 11 **Communications Applications of Polymers**, R. Spratling, British Telecom.

Report 12 **Process Control in the Plastics Industry**, R.F. Evans, Engelmann & Buckham Ancillaries.

Volume 2

Report 13 **Injection Moulding of Engineering Thermoplastics**, A.F. Whelan, London School of Polymer Technology.

Report 14 **Polymers and Their Uses in the Sports and Leisure Industries**, A.L. Cox and R.P. Brown, Rapra Technology Ltd.

Report 15 **Polyurethane, Materials, Processing and Applications**, G. Woods, Consultant.

Report 16 **Polyetheretherketone**, D.J. Kemmish, ICI, Wilton.

Report 17 **Extrusion**, G.M. Gale, Rapra Technology Ltd.

Report 18 **Agricultural and Horticultural Applications of Polymers**, J.C. Garnaud, International Committee for Plastics in Agriculture.

Report 19 **Recycling and Disposal of Plastics Packaging**, R.C. Fox, Plas/Tech Ltd.

Report 20 **Pultrusion**, L. Hollaway, University of Surrey.

Report 21 **Materials Handling in the Polymer Industry**, H. Hardy, Chronos Richardson Ltd.

Report 22 **Electronics Applications of Polymers**, M.T.Goosey, Plessey Research (Caswell) Ltd.

Report 23 **Offshore Applications of Polymers**, J.W.Brockbank, Avon Industrial Polymers Ltd.

Report 24 **Recent Developments in Materials for Food Packaging**, R.A. Roberts, Pira Packaging Division.

Volume 3

Report 25 **Foams and Blowing Agents**, J.M. Methven, Cellcom Technology Associates.

Report 26 **Polymers and Structural Composites in Civil Engineering**, L. Hollaway, University of Surrey.

Report 27 **Injection Moulding of Rubber**, M.A. Wheelans, Consultant.

Report 28 **Adhesives for Structural and Engineering Applications**, C. O'Reilly, Loctite (Ireland) Ltd.

Report 29 **Polymers in Marine Applications**, C.F.Britton, Corrosion Monitoring Consultancy.

Report 30 **Non-destructive Testing of Polymers**, W.N. Reynolds, National NDT Centre, Harwell.

Report 31 **Silicone Rubbers**, B.R. Trego and H.W.Winnan, Dow Corning Ltd.

Report 32 **Fluoroelastomers - Properties and Applications**, D. Cook and M. Lynn, 3M United Kingdom Plc and 3M Belgium SA.

Report 33 **Polyamides**, R.S. Williams and T. Daniels, T & N Technology Ltd. and BIP Chemicals Ltd.

Report 34 **Extrusion of Rubber**, J.G.A. Lovegrove, Nova Petrochemicals Inc.

Report 35 **Polymers in Household Electrical Goods**, D.Alvey, Hotpoint Ltd.

Report 36 **Developments in Additives to Meet Health and Environmental Concerns**, M.J. Forrest, Rapra Technology Ltd.

Volume 4

Report 37 **Polymers in Aerospace Applications**, W.W. Wright, University of Surrey.

Report 39 **Polymers in Chemically Resistant Applications**, D. Cattell, Cattell Consultancy Services.

Report 41 **Failure of Plastics**, S. Turner, Queen Mary College.

Report 42 **Polycarbonates**, R. Pakull, U. Grigo, D. Freitag, Bayer AG.

Report 43 **Polymeric Materials from Renewable Resources**, J.M. Methven, UMIST.

Report 44 **Flammability and Flame Retardants in Plastics**, J. Green, FMC Corp.

Report 45 **Composites - Tooling and Component Processing**, N.G. Brain, Tooltex.

Report 46 **Quality Today in Polymer Processing**, S.H. Coulson, J.A. Cousans, Exxon Chemical International Marketing.

Report 47 **Chemical Analysis of Polymers**, G. Lawson, Leicester Polytechnic.

Volume 5

Report 49 **Blends and Alloys of Engineering Thermoplastics**, H.T. van de Grampel, General Electric Plastics BV.

Report 50 **Automotive Applications of Polymers II**, A.N.A. Elliott, Consultant.

Report 51 **Biomedical Applications of Polymers**, C.G. Gebelein, Youngstown State University / Florida Atlantic University.

Report 52 **Polymer Supported Chemical Reactions**, P. Hodge, University of Manchester.

Report 53 **Weathering of Polymers**, S.M. Halliwell, Building Research Establishment.

Report 54 **Health and Safety in the Rubber Industry**, A.R. Nutt, Arnold Nutt & Co. and J. Wade.

Report 55 **Computer Modelling of Polymer Processing**, E. Andreassen, Å. Larsen and E.L. Hinrichsen, Senter for Industriforskning, Norway.

Report 56 **Plastics in High Temperature Applications**, J. Maxwell, Consultant.

Report 57 **Joining of Plastics**, K.W. Allen, City University.

Report 58 **Physical Testing of Rubber**, R.P. Brown, Rapra Technology Ltd.

Report 59 **Polyimides - Materials, Processing and Applications**, A.J. Kirby, Du Pont (U.K.) Ltd.

Report 60 **Physical Testing of Thermoplastics**, S.W. Hawley, Rapra Technology Ltd.

Volume 6

Report 61 **Food Contact Polymeric Materials**, J.A. Sidwell, Rapra Technology Ltd.

Report 62 **Coextrusion**, D. Djordjevic, Klöckner ER-WE-PA GmbH.

Report 63 **Conductive Polymers II**, R.H. Friend, University of Cambridge, Cavendish Laboratory.

Report 64 **Designing with Plastics**, P.R. Lewis, The Open University.

Report 65 **Decorating and Coating of Plastics**, P.J. Robinson, International Automotive Design.

Report 66 **Reinforced Thermoplastics - Composition, Processing and Applications**, P.G. Kelleher, New Jersey Polymer Extension Center at Stevens Institute of Technology.

Report 67 **Plastics in Thermal and Acoustic Building Insulation**, V.L. Kefford, MRM Engineering Consultancy.

Report 68 **Cure Assessment by Physical and Chemical Techniques**, B.G. Willoughby, Rapra Technology Ltd.

Report 69 **Toxicity of Plastics and Rubber in Fire**, P.J. Fardell, Building Research Establishment, Fire Research Station.

Report 70 **Acrylonitrile-Butadiene-Styrene Polymers**, M.E. Adams, D.J. Buckley, R.E. Colborn, W.P. England and D.N. Schissel, General Electric Corporate Research and Development Center.

Report 71 **Rotational Moulding,** R.J. Crawford, The Queen's University of Belfast.

Report 72 **Advances in Injection Moulding**, C.A. Maier, Econology Ltd.

Volume 7

Report 73 **Reactive Processing of Polymers**, M.W.R. Brown, P.D. Coates and A.F. Johnson, IRC in Polymer Science and Technology, University of Bradford.

Report 74 **Speciality Rubbers**, J.A. Brydson.

Report 75 **Plastics and the Environment**, I. Boustead, Boustead Consulting Ltd.

Report 76 **Polymeric Precursors for Ceramic Materials**, R.C.P. Cubbon.

Report 77 **Advances in Tyre Mechanics**, R.A. Ridha, M. Theves, Goodyear Technical Center.

Report 78 **PVC - Compounds, Processing and Applications**, J.Leadbitter, J.A. Day, J.L. Ryan, Hydro Polymers Ltd.

Report 79 **Rubber Compounding Ingredients - Need, Theory and Innovation, Part I: Vulcanising Systems, Antidegradants and Particulate Fillers for General Purpose Rubbers**, C. Hepburn, University of Ulster.

Report 80 **Anti-Corrosion Polymers: PEEK, PEKK and Other Polyaryls**, G. Pritchard, Kingston University.

Report 81 **Thermoplastic Elastomers - Properties and Applications**, J.A. Brydson.

Report 82 **Advances in Blow Moulding Process Optimization**, Andres Garcia-Rejon,Industrial Materials Institute, National Research Council Canada.

Report 83 **Molecular Weight Characterisation of Synthetic Polymers**, S.R. Holding and E. Meehan, Rapra Technology Ltd. and Polymer Laboratories Ltd.

Report 84 **Rheology and its Role in Plastics Processing**, P. Prentice, The Nottingham Trent University.

Volume 8

Report 85 **Ring Opening Polymerisation,** N. Spassky, Université Pierre et Marie Curie.

Report 86 **High Performance Engineering Plastics**, D.J. Kemmish, Victrex Ltd.

Report 87 **Rubber to Metal Bonding**, B.G. Crowther, Rapra Technology Ltd.

Report 88 **Plasticisers - Selection, Applications and Implications**, A.S. Wilson.

Report 89 **Polymer Membranes - Materials, Structures and Separation Performance**, T. deV. Naylor, The Smart Chemical Company.

Report 90 **Rubber Mixing**, P.R. Wood.

Report 91 **Recent Developments in Epoxy Resins**, I. Hamerton, University of Surrey.

Report 92 **Continuous Vulcanisation of Elastomer Profiles**, A. Hill, Meteor Gummiwerke.

Report 93 **Advances in Thermoforming**, J.L. Throne, Sherwood Technologies Inc.

Report 94 **Compressive Behaviour of Composites**, C. Soutis, Imperial College of Science, Technology and Medicine.

Report 95 **Thermal Analysis of Polymers**, M. P. Sepe, Dickten & Masch Manufacturing Co.

Report 96 **Polymeric Seals and Sealing Technology**, J.A. Hickman, St Clair (Polymers) Ltd.

Volume 9

Report 97 **Rubber Compounding Ingredients - Need, Theory and Innovation, Part II: Processing, Bonding, Fire Retardants**, C. Hepburn, University of Ulster.

Report 98 **Advances in Biodegradable Polymers**, G.F. Moore & S.M. Saunders, Rapra Technology Ltd.

Report 99 **Recycling of Rubber**, H.J. Manuel and W. Dierkes, Vredestein Rubber Recycling B.V.

Report 100 **Photoinitiated Polymerisation - Theory and Applications**, J.P. Fouassier, Ecole Nationale Supérieure de Chimie, Mulhouse.

Report 101 **Solvent-Free Adhesives**, T.E. Rolando, H.B. Fuller Company.

Report 102 **Plastics in Pressure Pipes**, T. Stafford, Rapra Technology Ltd.

Report 103 **Gas Assisted Moulding**, T.C. Pearson, Gas Injection Ltd.

Report 104 **Plastics Profile Extrusion**, R.J. Kent, Tangram Technology Ltd.

Report 105 **Rubber Extrusion Theory and Development**, B.G. Crowther.

Report 106 **Properties and Applications of Elastomeric Polysulfides**, T.C.P. Lee, Oxford Brookes University.

Report 107 **High Performance Polymer Fibres**, P.R. Lewis, The Open University.

Report 108 **Chemical Characterisation of Polyurethanes**, M.J. Forrest, Rapra Technology Ltd.

Volume 10

Report 109 **Rubber Injection Moulding - A Practical Guide**, J.A. Lindsay.

Report 110 **Long-Term and Accelerated Ageing Tests on Rubbers**, R.P. Brown, M.J. Forrest and G. Soulagnet, Rapra Technology Ltd.

Report 111 **Polymer Product Failure**, P.R. Lewis, The Open University.

Report 112 **Polystyrene - Synthesis, Production and Applications**, J.R. Wünsch, BASF AG.

Report 113 **Rubber-Modified Thermoplastics**, H. Keskkula, University of Texas at Austin.

Report 114 **Developments in Polyacetylene - Nanopolyacetylene**, V.M. Kobryanskii, Russian Academy of Sciences.

Report 115 **Metallocene-Catalysed Polymerisation**, W. Kaminsky, University of Hamburg.

Report 116 **Compounding in Co-rotating Twin-Screw Extruders**, Y. Wang, Tunghai University.

Report 117 **Rapid Prototyping, Tooling and Manufacturing**, R.J.M. Hague and P.E. Reeves, Edward Mackenzie Consulting.

Report 118 **Liquid Crystal Polymers - Synthesis, Properties and Applications**, D. Coates, CRL Ltd.

Report 119 **Rubbers in Contact with Food**, M.J. Forrest and J.A. Sidwell, Rapra Technology Ltd.

Report 120 **Electronics Applications of Polymers II**, M.T. Goosey, Shipley Ronal.

Volume 11

Report 121 **Polyamides as Engineering Thermoplastic Materials**, I.B. Page, BIP Ltd.

Report 122 **Flexible Packaging - Adhesives, Coatings and Processes**, T.E. Rolando, H.B. Fuller Company.

Report 123 **Polymer Blends**, L.A. Utracki, National Research Council Canada.

Report 124 **Sorting of Waste Plastics for Recycling**, R.D. Pascoe, University of Exeter.

Report 125 **Structural Studies of Polymers by Solution NMR**, H.N. Cheng, Hercules Incorporated.

Report 126 **Composites for Automotive Applications**, C.D. Rudd, University of Nottingham.

Report 127 **Polymers in Medical Applications**, B.J. Lambert and F.-W. Tang, Guidant Corp., and W.J. Rogers, Consultant.

Report 128 **Solid State NMR of Polymers**, P.A. Mirau, Lucent Technologies.

Report 129 **Failure of Polymer Products Due to Photo-oxidation**, D.C. Wright.

Report 130 **Failure of Polymer Products Due to Chemical Attack**, D.C. Wright.

Report 131 **Failure of Polymer Products Due to Thermo-oxidation**, D.C. Wright.

Report 132 **Stabilisers for Polyolefins**, C. Kröhnke and F. Werner, Clariant Huningue SA.

Volume 12

Report 133 **Advances in Automation for Plastics Injection Moulding**, J. Mallon, Yushin Inc.

Report 134 **Infrared and Raman Spectroscopy of Polymers**, J.L. Koenig, Case Western Reserve University.

Report 135 **Polymers in Sport and Leisure**, R.P. Brown.

Report 136 **Radiation Curing**, R.S. Davidson, DavRad Services.

Report 137 **Silicone Elastomers**, P. Jerschow, Wacker-Chemie GmbH.

Report 138 **Health and Safety in the Rubber Industry**, N. Chaiear, Khon Kaen University.

Report 139 **Rubber Analysis - Polymers, Compounds and Products**, M.J. Forrest, Rapra Technology Ltd.

Report 140 **Tyre Compounding for Improved Performance**, M.S. Evans, Kumho European Technical Centre.

Report 141 **Particulate Fillers for Polymers**, Professor R.N. Rothon, Rothon Consultants and Manchester Metropolitan University.

Report 142 **Blowing Agents for Polyurethane Foams**, S.N. Singh, Huntsman Polyurethanes.

Report 143 **Adhesion and Bonding to Polyolefins**, D.M. Brewis and I. Mathieson, Institute of Surface Science & Technology, Loughborough University.

Report 144 **Rubber Curing Systems**, R.N. Datta, Flexsys BV.

Volume 13

Report 145 **Multi-Material Injection Moulding**, V. Goodship and J.C. Love, The University of Warwick.

Report 146 **In-Mould Decoration of Plastics**, J.C. Love and V. Goodship, The University of Warwick

Report 147 **Rubber Product Failure**, R.P. Brown.

Plastics Waste - Feedstock Recycling, Chemical Recycling and Incineration

A. Tukker, TNO

ISBN 1-85957-331-2

Contents

1 Introduction

Recycling of packaging, or plastics materials in general, is a good thing for the environment. This opinion now has found firm ground all over the European Union, and various regulations have been implemented or are in development to ensure a reasonable recycling rate for waste streams such as end-of-life packaging, automotive waste and waste electrical and electronic products.

But here much of the consensus seems to stop. EU member states develop different recycling strategies. Debate is going on about how 'recycling' actually should be defined. Is, as some seem to advocate, energy recovery in environmental terms as acceptable as other recycling methods? Is mechanical recycling to be preferred above other possible forms of recycling? Is – in the packaging field – one-way packaging as acceptable as re-usable packaging? For packaging (where EU legislation has been in place for the longest time) the EU Packaging Directive has set quotas for the different forms of recovery and recycling, but the discussion pops up regularly on the agenda.

When we look at the types of packaging materials or indeed, waste components in general (glass, paper, metal, plastic), it is clear that for plastics this discussion is probably most prominent. Particularly compared to glass and metals it is relatively difficult to collect plastics and/or upgrade them into a clean mono-flow, which is a precondition for successful mechanical recycling. Alternative recycling methods like feedstock or chemical recycling are often proposed.

This Rapra report will discuss the options for feedstock recycling of plastics waste. Furthermore, the report will include a short discussion on the environmental and economical pros and cons of feedstock recycling in comparison to regular incineration of Municipal Solid Waste (MSW) and mechanical recycling. This will be done on the basis of a number of existing life-cycle-assessments. In this respect, this report benefits from the experience of the TNO-CML Centre of Chain Analysis. This Centre is the joint-venture on life-cycle-assessment (LCA) and substance flow analysis of the Centre of Environmental Science of Leiden University, and TNO, one of the biggest research institutes on technology and innovation in the EU. Examples of relevant studies on which this report was based include: chemical recycling of plastics waste (a.1), for the EU's Green Paper on PVC waste management; a review of a study of Oko-Institut on packaging waste for Duales System Deutschland (DSD); the TNO/CE 'Cold box' study for APME; and some major LCAs for the Dutch Packaging Covenant and industrial parties.

This report is structured as follows:

- Section 2 gives a general introduction to plastics waste management options;
- Section 3 discusses options for feedstock recycling of mixed plastic waste;
- Section 4 discusses options for feedstock recycling of PVC-rich waste;
- Section 5 discusses options for chemical recycling of specific plastics waste;
- Section 6 discusses some alternative options for mixed plastic waste (MPW) treatment;
- Section 7 gives a brief discussion of the pros and cons of feedstock recycling compared to other options.

2 Plastics Waste Recycling: An Overview

The draft EU Directive on Packaging waste contains definitions of various forms of recycling and reuse:

(1) Reuse implies the use of the same product without essential changes in a new use cycle (e.g., refillable packaging after cleaning);

(2) Material recycling implies the application of the material used, without changing the chemical structure, for a new application;

(3) Chemical recycling implies a change of the chemical structure of the material, but in such a way that the resulting chemicals can be used to produce the original material again;

(4) Feedstock recycling implies a change of the chemical structure of the material, where the resulting chemicals are used for another purpose than producing the original material;

(5) Recycling with energy recovery implies input into a device where the energy content of the input material is used.

The difference between chemical and feedstock recycling is peculiar. As we will see in the next sections, there is, in essence, hardly any technology that recycles polymers into its own monomers. In this report we will concentrate on feedstock recycling, but in this 'broad' definition we will include chemical recycling as well, see Section 5.

The above illustrates that various different types of feedstock recycling exist. **Figure 1** illustrates the material chain for these and other recycling routes, in relation to some relevant input criteria for each route. As explained later, processes dealing with MPW often have a limited chlorine tolerance, e.g., due to corrosion problems or the feedstock quality produced. In a study for Directorate General (DG) III of the EU on chemical recycling of plastics waste (a.1), we performed an extended survey of feedstock recycling processes. It appeared that the list of options broadly could be grouped into three main categories in view of the need to apply specific technologies for each type of plastic waste:

(1) Mixed plastic waste (MPW) with a limited chlorine (i.e., PVC) content;

(2) PVC-rich plastic waste;

(3) Specific plastic waste, particularly PET, PUR and nylon.

In the next three chapters, we will discuss technologies and initiatives relevant for MPW, PVC-rich waste and specific plastics waste. The technologies will be compared in terms of technical performance, costs per tonne of waste treated ('gate fee', which is the cost charged when a tonne of waste meeting the input criteria is delivered to the gate of the plant), environmental aspects and input criteria. The review concentrates on initiatives that are or have been operational or which are generally regarded to be major candidates to be in operation in the next 3-5 years. Hence, we will not discuss the many tests on laboratory scale or general ideas that can be found in literature, but which are unlikely to be realised in practice in the near future (see Appendix D in a.1).

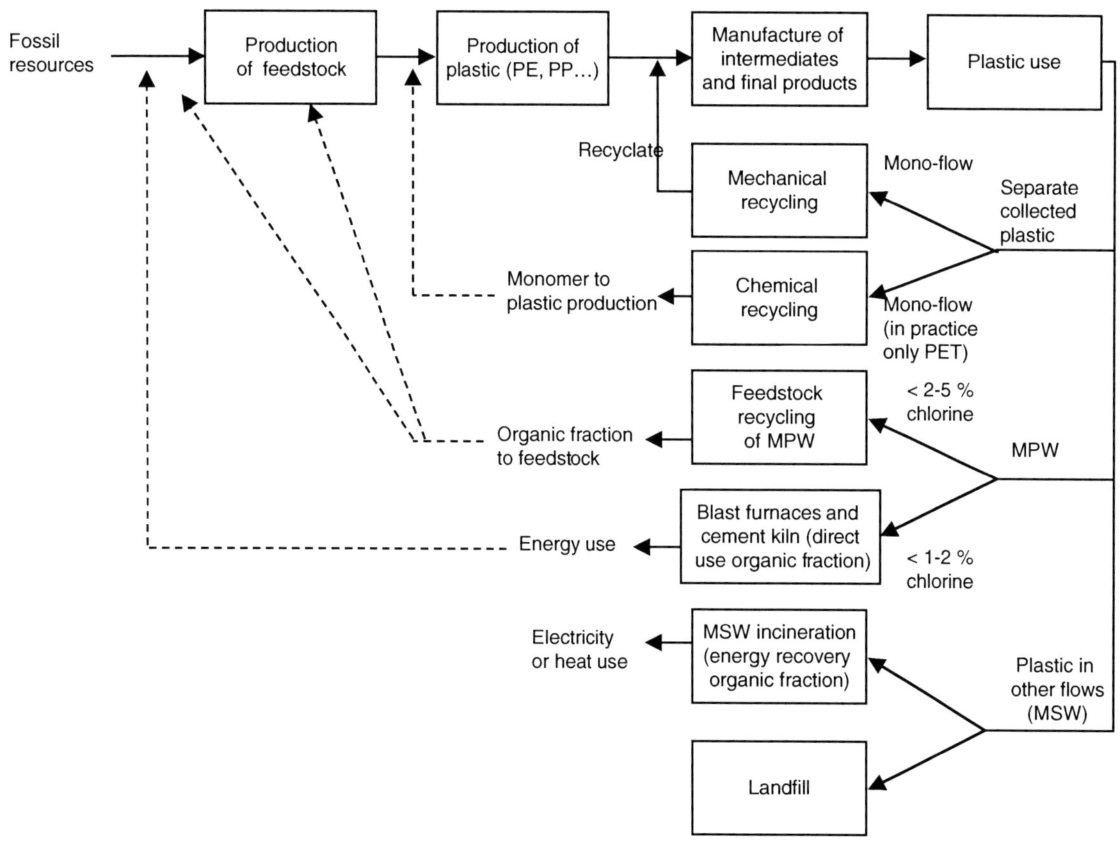

Figure 1

Schematic of material chains related to plastics, from production to waste disposal routes

3 Feedstock Recycling of Mixed Plastic Waste

3.1 Introduction

Concerning feedstock recycling of mixed plastic waste (MPW) with a low chlorine content, the following initiatives seem to be most promising. They are either operating in practice, have operated in the past, or have a fair chance of becoming operational in the short-term. Methods include:

(1) Texaco gasification process (Netherlands, pilot in the US)

(2) Polymer cracking process (consortium project, pilot)

(3) BASF conversion process (Germany, pilot but on hold)

(4) Use as a reducing agent in blast furnaces (Germany, operational): In this process MPW is used as a reducing agent, and hence this is generally seen as a form of feedstock recycling. For instance, in Germany this is one of the most important technologies by which the ambitious German recycling target for plastic packaging waste is met (a.2).

(5) Veba Combi Cracking process (Germany, operational until 2000)

(6) Pressurised fixed bed gasification from SVZ (Germany, operational)

These processes are discussed below.

3.2 Texaco Gasification Process

3.2.1 Background and Current Status

For over 40 years Texaco has been commercially involved in gasification. For regular feedstock, the Texaco gasification technology has proven its reliability and flexibility in over 100 plants throughout the world.

It took a long time before Texaco considered plastic waste as a potential feedstock. However, with the emergence of more stringent demands in waste management in general and plastics waste in particular, this situation changed in the mid nineties. Texaco is confident that its process is capable of dealing with plastics waste and that with some adaptations to its installation this can be an efficient and commercially viable feedstock. Therefore, Texaco started pilot plant experiments with mixed plastic waste (10 tonnes/day) in its plant in Montebello, California, USA (a.3).

Commercialisation to a full scale plant was considered by a Dutch-oriented consortium comprising Texaco, Air Products, Roteb and VAM (the latter two being Dutch waste management companies). The idea was to use plastics from the VAM mechanical separation plant for municipal and industrial solid waste as a feedstock in a plant based on the Texaco process to be built in Pernis, near Rotterdam in the Netherlands. The idea was that this would help to reach the rather stringent Dutch recycling quota for packaging waste without the need to collect the waste separately. However, in the end this initiative did not materialise. VAM, obviously, had a number of alternatives as an outlet for its plastics waste fraction – and with the opening of the EU borders for waste for recycling and recovery these included among others cement kilns abroad, but also energy power plants in the Netherlands. Though none of the parties officially commented on this, apparently VAM and Texaco could not agree on an attractive gate fee in combination with a certain supply of feedstock over time. Hence, the project currently is on hold.

3.2.2 Description of the Process

Texaco gasification is based on a combination of two process steps, a liquefaction step and an entrained bed gasifier. In the liquefaction step the plastic waste is cracked under relatively mild thermal conditions. This depolymerisation results in a synthetic heavy oil and a gas fraction, which in part is condensable. The non-condensable fraction is used as a fuel in the process. The process is very comparable to the cracking of vacuum residues that originate from oil recycling processes.

Particles are removed from the heavy fraction by filtration. The condensed gas fraction and the filtered heavy oil are then fed jointly into a gasifier, including chlorine containing gases which might arise from any PVC in the original MPW. The gasification takes places between 1200-1500 °C in the presence of steam and oxygen. The pressure maintained in the process depends on the application of the product (synthesis gas). Before the synthesis gas is fed into the new primary process, impurities like HCl and HF are removed in a number of cleaning steps. Chlorine present in the feedstock is captured by washing the raw syngas under addition of NH_3 and converted into saleable NH_4Cl (a.4). Sulfur from MPW is won back in a pure, saleable form. This makes the resulting synthesis gas, containing mainly CO and H_2 ready for use in other processes. Apart from CO and H_2 the syngas contains smaller amounts of CH_4, CO_2, H_2O and some inert gases.

Metals in the feedstock end up in slag and fines. The slag meets the quality standards of the Dutch Building decree, and the fines have a comparable quality to municipal solid waste incineration (MSWI) fly-ash (a.4).

In summary, treatment of MPW in this process leads to the following products:

- Synthesis gas. 150 tonnes of mixed plastics per day produces roughly 350,000 Nm^3 per day of clean synthesis gas (mainly H_2/CO) that can be used as feedstock in petrochemical processes.
- Pure sulfur.
- Saleable NH_4Cl.
- Vitrified slag. This has a quality that meets the requirements of Dutch legislation for secondary building materials.
- Fines. These have a quality that would match the quality of fly ash from Dutch municipal solid waste incineration plants.

3.2.3 Acceptance Criteria for the Input Material

Texaco has communicated the following acceptance criteria for its process (a.1). Depending on the design of the purification step after gasification, the process can deal well with up to 10% PVC in its feedstock. The tolerance to non-plastic materials like inorganics and paper is thought to be around 10%. Other acceptance criteria include:

- Material texture — Dry to the touch, not sticky, free flowing
- Physical description — Shredded or chipped
- Size — Less than 10 cm
- Physical fines content — Less than 1% under 250 μm
- Bulk density — > 100 g/litre
- Form at delivery — baled or agglomerated
- Plastics content — > 90 wt%
- Free metals — < 1 wt%
- PVC content — < 10 wt%
- Ash content — < 6 wt%
- Residual moisture — < 5 wt%
- Paper content — < 10 wt%

3.2.4 Environmental and Cost Performance

Croezen and Sas (a.4) have published an extensive LCA for treatment of MPW with the Texaco process. There is a detailed discussion on inputs and emissions there. No specific problems with emissions control were mentioned by these authors. As for cost performance, detailed data of the cost structure have not been made public. The general figure circulating for the likely gate fee of the Texaco process is €90 to €135 per tonne for a 50 ktonne/year plant, decreasing to €50/tonne for a 200 ktonne/year plant.

3.3 The Polymer Cracking Process (Consortium Project)

3.3.1 Background and Current Status

BP Chemicals is another company that has a cracking process available that could be of use for feedstock recycling of plastics waste. Also here the challenge of plastics recycling that emerged after 1990 led to the formation of a consortium of interested industries, aiming at developing the technology. Initially the group included Elf Atochem, DSM, Fina and EniChem. At the time of the successful trials of 1997 the consortium consisted of BP Chemicals, Elf Atochem, EniChem, DSM, CREED and the Association of Plastic Manufacturers in Europe (APME). In view of demands for recycling of plastics waste APME supported the research and development phase of the project.

The 'Polymer Cracking Process' is basically a fluid bed cracking process. It was first tested on lab scale around 1990. Subsequent improvements and research lead to a successful demonstration of treatment of MPW at continuous pilot plant scale at BP's Grangemouth site. The pilot plant has a nominal 400 tonne/year capacity. However, it runs continuously on a 50 kg/h scale as it has limited product storage. The next steps in the development process include modifications which allow for optimisation and above all scale-up. Plans for realising such a full scale plant have been developed for the UK in view of its shortage of plastic recycling capacity. However, until now it was not yet possible to realise the right partnerships and economic factors that allowed for full commercialisation.

3.3.2 Description of the Process

Before MPW is fed into the process, a basic separation of the non-plastic fraction and size reduction is needed. This prepared feedstock is then introduced in the heated fluidised bed reactor which forms the core of the process. The reactor operates at approximately 500 °C in the absence of air. At this temperature, thermal cracking of the plastics occurs. The resulting hydrocarbons vapourise and leave the bed with the fluidising gas. Solid particles, mainly impurities formed from, e.g., stabilisers in plastics, as well as some coke formed in the process mainly accumulate in the bed. Another fraction is blown out with the hot gas and captured in a cyclone.

Any chlorine in the feedstock (e.g., from PVC in the MPW) is converted into HCl, and the gas is purified with lime. The main result is a $CaCl_2$-fraction that has to be landfilled.

By cooling the gas is condensed and then available as hydrocarbon feedstock for other processes (some 85% of the MPW input). The light hydrocarbon gas (15% of the MPW input) that remains after cooling is compressed, reheated and returned to the reactor as fluidising gas. It can also be used as a fuel for the cracking process, though other recovery options are being studied as well.

The gas has a high content of monomers (ethylene and propylene) and other useful hydrocarbons with only some 15% being methane. The feedstock is collected in two stages since the heavy fraction is a wax below about 60 °C. The heavy fraction is typically 60% by weight of the product with the light fraction being 40% by weight.

Impurities like chlorine are effectively re-used. At an input of 1% of chlorine in the MPW (2% PVC), the products will contain around 10 ppm Cl. This is somewhat higher than the specifications of 5 ppm typical for refinery use. However, in view of the high dilution likely in any refinery or petrochemical application, BP assumes that this is acceptable (a.5). Also, metals like Pb, Cd and Sb can be removed to very low levels in the products. Tests have shown that all the hydrocarbon products can be used further in refineries.

3.3.3 Acceptance Criteria for the Input Material

Typical input specifications for BP's Grangemouth pilot plant are:

- Polyolefins: 80 (min. 70) wt%
- Polystyrene: 15 (max. 30) wt%
- PET: 3 (max. 5) wt%
- PVC: 2 (max. 4) wt%
- Total plastic content: 95 (min. 90) wt%
- Ash: 2 (max. 5) wt%
- Moisture: 0.5 (max. 1) wt%
- Metal pieces: max. 1 wt%
- Size: 1-20 mm
- Fines sub-250 micron: max. 1 wt%
- Bulk density: 400 (min. 300) kg/m^3

Plans exist to test other materials to examine further potential on other non-packaging feed supplies. This might extend the limits of this specification. It is said that the process would tolerate short-term excursions of higher chlorine content, e.g., 5% wt Cl.

3.3.4 Environmental and Economic Performance

It is difficult to give precise data at this time as the process is still in the development stage. In principle, in terms of heating, the process can be self-sufficient. However, depending on input quality of the MPW some additional gas might be needed to run the process. Other inputs into the process are:

- electric power approximately 60 kW/tonne feed plastic
- cooling water 40 m^3/tonne feed plastic
- steam 1.2 tonne/tonne feed

All emissions will be very low and will comply with local regulations. Waste products are about 0.2 kg/kg of total solids feed. Note, this total solids feed includes both feed plastic and the solids used as make-up in the process.

The cost of treatment to process one tonne is difficult to define since it depends on many factors such as scale, location, scope, preparation stages, and economic parameters used. As a rough example, for a 25,000 tpa plant in Western Europe (1998 prices), BP estimated the investment to be £15 to £20 million. This would imply a gate fee of around £172 per tonne (some €250). For a 50,000 tpa plant the gate fee could be £100 per tonne (some €150).These figures are net, i.e., include product values but exclude collection and preparation.

3.4 The BASF Conversion Process

3.4.1 Background and Current Status

The Duales System Deutschland (DSD) gave a boost to the development of packaging waste recycling initiatives in Germany. One of the initiatives for plastics waste was the BASF feedstock recycling process. It was designed to handle the recycling of mixed plastic waste supplied by the DSD collection. A 15,000 tpa pilot plant was operational in Ludwigshafen in 1994. From that moment on, a discussion between DSD and BASF took place on the prospects of setting up a full scale plant. Initially, DSD estimated a need for feedstock recycling in Germany of 750,000 tpa. In 1995, this estimate was already reduced to 400,000 tpa. In this period, BASF kept the pilot plant operational to gain experience with the process and to allow further developments. However, in 1996 BASF announced, after consultation with DSD and the Deutsche Kunststoff Recycling AG (DKR), that it would close its pilot plant. Officially no reasons have been disclosed, but it seems likely that also here a classical problem occurred: before making an investment in a major new industrial installation, certainty is needed about a price level and volume of the feedstock for a reasonable time period – and apparently such a long-term commitment was not possible.

3.4.2 Description of the Process

The BASF process requires pretreatment. Plastics are separated from non-plastics, ground and agglomerated. The agglomerate is fed into the process.

In the first step, the plastic is melted and dehalogenated. Again, due to the presence of PVC in MPW this is one of the main impurities. The dehalogenation prevents corrosion problems from occurring in the rest of the installation. An advantage of the BASF process is that the chlorine is recovered as HCl, which can be used in other applications. A small fraction of the chlorine ends up as NaCl or $CaCl_2$ in an aqueous effluent (a.6). The result from this step is liquefied plastics and a gaseous fraction. The gaseous fraction can be used as feedstock in a cracker after compression.

The liquefied plastic fraction is heated to over 400 °C. This leads to cracking of the plastic into components of different chain lengths. Gases count for 20%-30% and oils for 60%-70%; they are separated by distillation. Any naphtha produced is treated in a steam cracker, resulting in monomers like ethylene and propylene that are recovered. Such monomers can be used to produce plastics again. The heavy fractions can be processed into synthesis gas or conversion coke and then be transferred for further use. At most 5% of the input is converted into a mineral fraction. It is likely that this consists mainly of the inorganic additives in plastics.

In sum, the process results into the following products:

- HCl, which is for a small part neutralised or processed in a hydrochloric acid production plant;
- naphtha to be treated in a steam cracker;
- monomers, e.g., ethylene and propylene, which can be used for the production of virgin plastic materials;
- high boiling oils, which can be processed into synthesis gas or conversion coke and then transferred for further use;
- residues.

3.4.3 Environmental and Economic Performance

The process is fairly robust. As for one of the main problematic inputs, chlorine from PVC, the plant was able to handle MPW with the regularly occurring PVC content of 4%-5%. This implies a chlorine tolerance of some 2.5%. As for the emissions and resource use, Heyde and Kremer (a.6) have carried out an extensive study. All emissions will comply with local regulations.

As for processing costs, the BASF process would require a gate fee of €160 per tonne for a 300,000 tonnes/year plant and a fee of €250 per tonne for a 150,000 tonnes/year plant. To our knowledge BASF has not disclosed a more detailed cost structure.

3.5 Use of Mixed Plastic Waste in Blast Furnaces

3.5.1 Background and Current Status

During steel production iron ore (Fe_2O_3) has to be reduced to metallic iron (Fe). This process takes place in a blast furnace where the iron ore is reduced with materials like coke, coal and/or heavy oil. Several steel producers are experimenting with replacing these primary materials with plastics waste. Important pioneers include British Steel (UK) and Stahlwerke Bremen, Germany. The latter company is the only one in the EU who now uses MPW as a reducing agent on a regular basis. Stahlwerke Bremen operates two blast furnaces to produce over 7,000 tonnes/day, or some 3 million tpa of pig iron. They started to investigate this possibility in 1993. Experiments started in February 1994 with a capacity of 50 tonnes/day of plastic waste. In mid 1995 a large size operation was started using agglomerated DSD waste (75,000 tonnes/year). The capacity was increased to 162,500 tpa MPW in 1998, which was some 25% of the recycling capacity for MPW in Germany (a.2). With SVZ (see Section 3.7) Stahlwerke Bremen is the only operational full scale treatment option for MPW via feedstock recycling in the EU.

The specific German regulation on packaging waste asks for high recycling targets and this might have been one of the reasons why this form of feedstock recycling is applied only in Germany. Though cost benefits are involved, other producers might be afraid of a lower reliability of their blast furnace operation in the learning phase – which can be a costly affair indeed. Just imagine a shut-down of a blast furnace of just a few days due to problems with this new type of reducing agent. However, if this hurdle can be overcome the potential capacity in the EU is huge. The total pig iron production in the EU is some 90 million tonnes, or some 30 times the capacity of Bremen Stahlwerke. This would imply a capacity of 5 million tonnes MPW per year for all European steel works.

3.5.2 Description of the Process

As indicated, pig iron production requires input of a reducing agent. Stahlwerke Bremen uses plastic waste as a substitute for fuel oil. Plastics are injected into the blast furnace in a similar way to coal powder or fuel oil. In order to remove fibres and metal particles a separation takes place. Large particles are separated via a screen of > 18 mm. The smaller plastic waste particles (< 18 mm) go to the injection vessel. There, an injection pressure of about 0.5 MPa is built up. Via a pneumatic process the plastics can be dosed and discharged into the blast furnace. The bulk density of the plastics has to be 0.3 tonnes/m^3.

As indicated, the reliability of the process is crucial and hence the input has to be controlled. MPW has the advantage that it contains relatively low amounts of sulfur, but the chlorine content has to be limited.

Concern has been expressed about the possible formation of dioxins and furans. However, measurements during experiments indicated that the emissions of dioxins and furans were not significantly elevated. Dioxin emissions with or without plastic input appeared to be about a factor of 100 below the standard of 0.1 ng/Nm^3 TEQ TCCD (toxicity equivalent in relation to the toxic dioxin TCCD) (a.7). This might be due to the benefit of the strongly reducing atmosphere and the high temperature of 2100 °C. In total, until now the conclusion has been that at current PVC levels in MSW, pretreatment for chlorine removal is unnecessary.

However, the PVC throughput in the blast furnace kiln is just a fraction of the total material throughput. This is comparable to MSWIs, where PVC in general forms less than 1% of the input. Under such circumstances, the relation between PVC input and dioxin formation appears quite difficult to asses. Most research reports claim that there is no clear relation (e.g., a.8, a.9). However, Greenpeace has published a number of reports that suggest otherwise (e.g., a.10). Furthermore, it has to be noted that the off-gas of blast furnaces is generally used as an energy carrier in other processes. Checks on dioxin formation are desirable there as well.

On top of this, PVC is by no means the only chlorine source. Other raw materials and (particularly for blast furnaces close to the sea) even the air used in incineration processes may have significant contributions to the chlorine throughput too.

3.5.3 Acceptance Criteria for the Input Material

The permit allows Stahlwerke Bremen to use 500 tonne MPW per day with a chlorine content of up to 1.5% (= ca. 3% PVC) on a daily average. This level seems to be a balance between the need to allow for a reasonable PVC tolerance in MPW (lower values are rare in MPW), and the desire of Bremen Stahlwerke to use a material that is as 'free' of impurities as possible. After all, chlorine has no added value in the process, and may only contribute to problems like corrosion in the blast furnace, etc. In sum, the 1.5% level seems to be a balance between commercial reality and a technical ideal.

3.5.4 Environmental and Economic Performance

The LCA of Heyde and Kremer (a.6) gives an extensive review of emissions and resource use. However, by and large one could assume that the emissions by using plastics as reducing agent will be more or less equal to the emissions that would occur if another reducing agent were to be used. For a discussion on the (probably limited) relevance of dioxin emission see Section 3.5.3.

As for economic data, Bremen Stahlwerke has not given any insight as to its gate fee. However, various sources indicated that Duales System Deutschland provides a cost contribution of about €100 per tonne. This is logical, since some adaptations in terms of feed preparation were needed. Furthermore, particularly initially, investments and investigations were needed to analyse if MPW could be introduced without problems. It is clear that this process has a big advantage over others, i.e., that the capital costs are low or negligible. Hence, it might well be that the €100 per tonne presumably paid by DSD is far above the cost price. The real marginal costs for Bremen Stahlwerke probably are much lower, and might be even negative (i.e., using plastic waste instead of coal forms a net financial gain). However, the practical gate fee will be established under the influence of market forces. The actual gate fee thus will mainly depend on the availability and the price of competing technologies for the treatment of plastic waste.

3.6 Veba Combi Cracking Process

3.6.1 Introduction

In 1981 Veba Oel started a hydrogenation plant for coal, which produced naphtha and gas oil. This plant is known as the Kohleöl Anlage Bottrop (KAB) in Germany. Veba modified the plant in 1987. With the Veba Combi Cracking (VCC) technology vacuum distillation residues of crude oil could be transformed into synthetic crude containing naphtha, gas oil and heavy distillates. Over time, Veba started to substitute its normal feedstocks by waste (among others, oil containing PCBs). Veba changed the plant again in 1992/1993 by adding a depolymerisation unit at the front of the process. This was done to allow for processing MPW collected via the DSD system. The capacity is about 10 tonnes per hour. In continuous operation (8,000 h/year), this implies a depolymerisation capacity of around 80,000 tpa. This level was indeed treated in 1998. However, DSD and Veba agreed to terminate the original contract for MPW treatment by the end of 1999 (whereas it was meant to continue to 2003). Since in the meantime the plant had only been processing DSD waste, Veba decided to close down the plant entirely.

Formally no explanations have been given for these decisions. However, it is widely believed that the Veba process could not compete economically with the SVZ and blast furnace processes.

3.6.2 Description of the Process

As indicated above, the plant consists of a VCC part and a depolymerisation part. Depolymerisation allows for further processing of the residues in the VCC section. The depolymerisation takes place between 350-400 °C. Here, at the same time chlorine is released. Over 80% of the chlorine input will become available as HCl in the light fraction and washed out in a purification process yielding technical HCl.

The gaseous product of the depolymerisation is partially condensed. The condensate, containing 18% of the chlorine input, is fed into a hydrotreater. The HCl is eliminated with the formation of water. The resulting Cl-free condensate and gas are mixed with the depolymerisate for treatment in the VCC section.

The depolymerisate is hydrogenated in the VCC section at 400-450 °C. This takes place under high pressure (about 10 MPa) in a liquid phase reactor. After separation and treatment in a fixed-bed hydrotreater a synthetic crude oil is formed, a valuable product which may be processed in any refinery. The separation also gives also a hydrogenated residue stream. This fraction contains the heavy hydrocarbons contaminated with ashes, metals and inert salts. This by-product is called hydrogenation bitumen and blended with the coal for coke production (2 wt%). Inorganic materials in the input most probably end up in this residue flow.

The light cracking products form a gas, which is cleaned from H_2S, HCl and ammonia. In total, only some 2% of the chlorine input is bound to CaO to form $CaCl_2$ in the process (a.6, a.11).

In sum, the following products are produced from MPW via this process:

- HCl
- syncrude from the VCC section; this liquid product is free of chlorine and low in oxygen and nitrogen
- hydrogenated solid residue, which can be blended with coal for coke production
- off-gas.

3.6.3 Acceptance Criteria for the Input Material

When the VEBA plant was operational, it had the following input specifications for the depolymerisation section:

- particle size < 1.0 cm
- bulk density ≥ 300 kg/m^3
- water content < 1.0 wt%
- PVC < 4% (≤ 2 wt% chlorine). Some have claimed that this process could deal with a PVC content of up to 10%. However, the firm repeatedly confirmed 2% chlorine or 4% PVC as the regular maximum.
- inerts < 4.5 wt% at 650 °C
- metal content < 1.0 wt%
- content of plastic ≥ 90.0 wt%.

As indicated, the plant treated MPW on a regular basis. However, successful tests have been done with electrical and electronic (E&E) waste as well. In the test, some 50 tons of E&E waste were mixed with some 250 tons of DSD waste (a.12).

3.6.4 Environmental and Economic Performance

A LCA on treatment options of MPW was performed by the Dutch Centrum voor Energiebesparing en Schone Technologie (CE, Delft) in 1994. This LCA used the VEBA process as an example for feedstock recycling (a.11). Another LCA was performed by Heyde and Kremer (a.6). Particularly the CE studies suggested that the VEBA process was a bit less advantageous than the Texaco process, mainly due to the fact that the Texaco process does not need agglomeration of MPW as pre-treatment, whereas the VEBA process apparently does.

As is the case for most other processes reviewed here, no detailed cost data are given by the firm that operates the process. It seems that the gate fee is around 250 Euro per tonne (compare also a.13).

3.7 SVZ Gasification Process

3.7.1 Background and Current Status

'Schwarze Pumpe', which is the informal name of the Sekundärrohstoff Verwertungs Zentrum (SVZ) operates a plant that converts several feedstocks into synthesis gas, methanol and electricity. SVZ was originally a coal gasification plant. However, some major investments made it possible to allow also waste materials, including plastics, as an input. The plant is currently fully operational. Waste and material that are accepted include contaminated wood, waste water purification sludge (including industrial sludges), waste derived fuel from MSW, paper fractions, plastic fractions, the light fraction of shredder waste, and liquid organic waste that arises from SVZ-related plants. The plant can treat about 410,000 tpa solid and 50,000 tpa liquid material. In the short-term, the plant will have a capacity for plastics of 140,000 tpa. In 1998 it treated already some 100,000 tpa plastics under contract from DSD. It is likely that the actual amount of plastics is higher, since other waste fractions treated contain plastics as well.

3.7.2 Description of the Process

Lignite, waste oil and MPW is fed into a reactor (a solid bed gasification kiln). Gasification is supported with oxygen and steam. These materials are fed in counter flow with the input materials. Like many of the processes discussed before, this results in hydrogen and CO (synthesis gas), liquid hydrocarbons and effluent. The liquid hydrocarbons are gasified. The resulting gases and the gases from the fluidised bed reactor are purified by the rectisol process, which removes components like H_2S and organic sulfur compounds.

The synthesis gas is mainly used for the production of methanol (70%). Another part (20%) is used for electricity production. Waste gas products are incinerated; the fate of any chlorine is not clear from the various descriptions available. Inorganic materials are converted into a slag, with low leaching characteristics (landfill class 1 according to the German TA Siedlungsabfall).

3.7.3 Acceptance Criteria for the Input Material

From the description above it already can be deduced that this plant can deal with a lot of waste types and hence is rather robust in terms of acceptance criteria. SVZ has experience with treating mixed plastics waste, waste derived fuel (a mixture of plastics, wood and paper), the shredder light fraction of car wrecks, and the plastic fraction from shredded white goods and electronics. SVZ can handle on average 2% chlorine in MPW, with short-term excursions to 6%. The overall chlorine content has to be controlled by a right blend with other waste types. SVZ does not favour a high chlorine input, due to problems like a higher risk of corrosion, and the need for neutralisation, leading to a salt that has to be landfilled at high cost. Some of the main acceptance criteria for MPW are:

- Particle size: > 20 to 80 mm
- Chlorine content: 2% as default, though higher concentrations are tolerable
- Ash content: up to 10% or more
- Calorific value: not critical.

3.7.4 Environmental and Economic Performance

A study of the Frauenhofer Institut Verfahrenstechnik und Verpackung (IVV) gave a good overview of the inputs and outputs of the SVZ process. **Table 1** reviews these data, see also the original study of Heyde and Kremer (a.6).

As is the case with most other owners of a feedstock recycling plant, SVZ has not given detailed insight into its cost structure nor the gate fee it obtains under its DSD contracts. However, indirectly one can deduce that SVZ's position is rather competitive compared to other chemical recycling initiatives. With the recent closure of VEBA, SVZ remains the only major chemical recycling plant that sustains the competition with as cost-effective options as treatment in steelworks. Hence, it seems unlikely that SVZ's gate fee will be much higher than the 100 Euro per tonne of MPW that seems to be valid for steelworks.

Table 1 Inputs and outputs of the SVZ process (based on Heyde and Kremer (a.6))

Inputs		Outputs	
MPW-agglomerate	763 g	Methanol	712 g
Waste oil	256 g	Synthesis gas	204 g
Lignite	1.25 kg	Electricity	2.28 MJ
Water	7.9 l	CO_2	6.32 kg
Oxygen	1.47 kg	Water vapour	9.9 kg
Fuel oil	40 g	Effluent	9.9 kg
Natural gas	0.1 m^2	Gypsum	0.1 kg
		Slag	0.9 g

4 Feedstock Recycling of PVC-rich Waste

4.1 Introduction

As indicated in the former chapter most processes for feedstock recycling have limitations with regard to the maximum allowable chlorine input. This makes those processes problematic for the treatment of PVC-rich MPW. Hence, a number of initiatives has been developed for the treatment of PVC waste. All these processes seek to recover the chlorine present in PVC in a usable form (HCl or a saleable chloride salt). The processes include:

(1) BSL incineration process (Germany);

(2) Akzo Nobel steam gasification process (Netherlands);

(3) Linde gasification process (Germany);

(4) NKT pyrolysis process (Denmark).

An important driver for the development of these processes is the pressure on PVC producers in various EU member states, particularly in the area of waste management. In response, the European PVC industry has promised to establish recycling schemes (including recycling targets) for typical products consisting largely of PVC, such as vinyl flooring, window frames, PVC pipes, etc. In this context, the European PVC industry is actively supporting the development of several recycling processes for PVC, including the Linde process and the Vinyloop® process discussed in the next chapter.

4.2 BSL Incineration Process

4.2.1 Background and Current Status

BSL Olefinverbund GmbH (80% DOW, 20% BvS) in Schkopau has built an incineration plant for chlorine-containing fluid and solid waste streams. The plant is meant as a general plant for treatment of organochlorine waste from among others the plants of BSL and DOW. The idea is to convert the chlorine to HCl in the incineration process, and to convert the organic matrix into harmless molecules like CO_2 and H_2O using the energy from the process itself. The HCl produced will be used by BSL Schkopau in other processes, most notably membrane electrolysis for chlorine production.

The plant capacity for all waste types together will be around 45,000 tpa. It will have a heat production capacity of 25 MW at 7,500 productive hours a year. The plant can also accept PVC. For such solid chlorinated waste at most 15,000 tpa/yr capacity is available. This has been based on a test with a DOW kiln in Stade, Germany. The BSL plant has been operational since mid 1999.

4.2.2 Description of the Process

The process is basically a rotary kiln design. Waste is first pretreated and then inserted in the rotary kiln, where it is incinerated with air. The chlorinated hydrocarbons are converted into H_2O, CO_2 and HCl. After that, in a wet scrubber the HCl is recovered as aqueous HCl. If needs be, natural gas or liquid energy carriers can be added in order to reach the necessary high temperatures in the afterburner.

In the post-combustion chamber temperatures of 900 °C to 1200 °C are reached. The kiln can – like any rotary kiln – handle solid, fluid and gaseous waste streams. Based on the heat capacity of the waste, halogen content, and potential slag formation, an optimal mixture of wastes is determined. By choosing the feed carefully, production of high-quality HCl can be assured. Furthermore, in this way a minimum formation of dioxins and furans can be ensured.

The flue gas from post-combustion is cooled from 1200 °C to 230 °C to 300 °C. Here, energy is recovered. Steam is produced that is added to the steam network of the BSL Schkopau site. In the flue gas purification, the HCl is absorbed from the flue gas by water. Also, other impurities are removed from the gas. The raw HCl is then purified to a useful feedstock.

The composition of the inert fraction depends on the chemical composition of the waste input. Most probably the inorganic parts of the PVC formulations, like metal-containing stabilisers, will end up in this slag.

In sum, the secondary products from this process are:

- High quality HCl, which can be used in several production processes;
- Steam;
- Inert slag.

4.2.3 Acceptance Criteria for the Input Material

The rotary kiln design allows for accepting a mix of high-chlorinated wastes (solvents, chlorinated tars, plastics). Such kilns are usually designed in relation to a specific optimal calorific value in the input. The input mix should be set in such a way that this optimal composition is approached (e.g., PVC waste and other waste streams with a lower calorific value). It is likely that a 100% input of PVC would lead to all kind of problems of temperature control due to its relatively high calorific value. Chlorine contents of over 50% can easily be accepted. A final demand is that the particle size should be 10 x 10 x 10 cm at maximum. This implies that sometimes waste has to be shredded before it can be put into the kiln. Other acceptance criteria have not been published in literature.

4.2.4 Environmental and Economic Performance

Emissions will have to meet the stringent German TA Luft standards. According to BSL, even lower values will be reached during normal operations. Waste water is treated at the central treatment plant of the Schkopau site. Waste outputs include slags and filter residues. The latter is for a large part fed again into the kiln, the remainder being landfilled as hazardous waste. The slag is inert and usually applied as a filler in old salt mines.

According to Pohle (a.13), the gate fee for a 250,000 tpa plant using a similar technology would be €250 per tonne. Informal information from circles within the PVC industry suggest even higher gate fees for PVC (up to €350 to €500).

4.3 Akzo Nobel Steam Gasification Process

4.3.1 Background and Current Status

Until recently, Akzo Nobel was, via its joint venture with Shell – Rovin, the main Dutch producer of chlorine, vinyl chloride, and PVC. This business recently has been taken over by the Japanese PVC producer Shin Etzu. They started to study a process for feedstock recycling of mixed plastic waste containing PVC in 1992. An initial investigation led to a preference for fast pyrolysis technology in a circulating fluid bed reactor system. This technique was developed originally for biomass gasification by Battelle, Columbia, USA. Following this choice, Akzo Nobel did some small-scale pilot plant tests (20-30 kg/h) with PVC cable and pipe scrap. Later, tests with mixed plastic waste on a larger scale (200-400 kg/h) were carried out with the support of the European Council of Vinyl Manufacturers (ECVM). The results were promising.

The next step would be to build a large scale plant of 50 ktonnes per year. However, such an investment needs financing and a commitment of the European PVC industry to this initiative (i.e., a choice for this technology as the feedstock recycling process for PVC waste). Building the plant would take about 5 years. At this stage, it seems that within the PVC industry there is more support for the Linde and Vinyloop® processes.

4.3.2 Description of the Process

The process is centred around two separate circulating fluid bed (CFB) reactors. They both operate at atmospheric pressure:

(1) The first reactor is a gasification (or fast pyrolysis) reactor in which PVC-rich waste is converted at 700-900 °C with steam into a gaseous product (fuel gas and HCl) and residual tar.

(2) The second reactor is a combustor that burns the residual tar to provide the heat for gasification.

Heat is transferred between the two reactors by circulating sand between the gasifier and combustor. Both reactors are of the riser type with a very short residence time. They allow for a high PVC waste throughput. The gasifiers have a reducing atmosphere, which prevents the formation of dioxins.

If tars are formed (which happened in the trial with mixed PVC waste), it is necessary to apply a partial oxidation to convert these tars into gaseous products. Via a quench, HCl is recovered from the product stream which consists of fuel gas and HCl. After purification the HCl can be used for producing ethylene dichloride (EDC) via oxychlorination. The recovery of Cl_2 is more then 90%, in general 94%-97%. Inert materials in the feed, such as the chalk and metal stabilisers present in a PVC-formulation, are separated from the flue gas as fly ash or will be set free as a bleed from the circulating sand.

The final output of the reactor is a synthesis gas. The composition will depend on the input. If much polypropylene (PP) and polyethylene (PE) is present, relatively high amounts of ethylene and propylene will be formed. If mainly PVC is present, HCl and CH_4

will be more dominant. In either case CO and H_2 will be the main components.

4.3.3 Acceptance Criteria for The Input Material

The testing does not yet allow for setting specific input criteria. However, the testing made clear that the process probably can handle a broad spectrum of materials, such as wood, biomass, mixed plastic and pure PVC waste. For instance tests have been done on PVC waste but also with a mixture of PVC, PE, other polymers, Cu, Al, chalk, cement and fibres.

4.3.4 Environmental and Economic Performance

The exact need for resources and the emissions will depend in part on the input of the process. On the basis of the tests thus far, for a waste containing 40% PVC and 25% inorganic fillers, the data shown in **Table 2** can be estimated (amounts per tonne of PVC).

As for a gate fee, estimates are difficult to give. Costs are dominated by the investment costs, which Akzo estimates at €25 million for a 50,000 tpa plant. There exists a large uncertainty in this estimate. For instance, if the feed also included brominated or fluorinated compounds, extra cleaning steps would be needed which would greatly enhance the investment. It is simply too early to give a good estimate of costs per tonne for treatment.

4.4 Linde Gasification Process

4.4.1 Background and Current Status

Linde KCA in Germany offers another technology for feedstock recycling that has been developed on the basis of a gasification method originally developed for lignite and coal. Linde's technology is based on gasification of waste in a slag bath and made suitable for PVC with the following in mind:

(1) Conversion of the chlorine in PVC into HCl, usable in oxychlorination for the production of EDC;

(2) Recovery of the chemically bound energy in the waste;

(3) Production of unavoidable waste products that can be disposed of in compliance with regulations.

In 1999, the European Council of Vinyl Manufacturers (ECVM) chose this process as the most robust and economical, and started to support it within the context of ECVM's recycling efforts. ECVM committed €3 million as a support to the building of a pilot plant. The pilot will be build at Solvay's Tavaux plant, located in the eastern part of France. If the pilot is successful, most probably a 25,000 tpa full scale plant will be build. However, it is not expected that such a full scale plant will be operational before 2005.

4.4.2 Description of the Process

The plastic waste is separated from steel and non-ferrous metals after crushing. The crushed plastics, sand, steam and oxygen are entered into the reactor. This is a pressurised reactor filled with slag, mainly consisting of silicates. The temperature is about 1400-1600 °C. The process is exothermic and the atmosphere reducing. The main products are a synthesis gas (CO/H_2) containing HCl and a slag. The latter most probably contains most of the metal stabilisers present in the PVC-formulation. HCl is washed from the synthesis gas with water and subsequently purified from heavy metals, chlorides and other halogens. Via distillation of HCl pure HCl gas is produced. The clean synthesis

Table 2 Inputs and outputs for 40% PVC/25% inorganic fillers waste processing in the Akzo Nobel steam gasification process per tonne of PVC

Inputs		Outputs	
Steam	0.3 t	HCl	0.21 t
Process water	1.0 m^3	Synthesis gas	0.9 t
Air	2.3 t	Fly ash and bottom ash	0.22 t (to be disposed of)
Electricity	115,200 kWh		
Cooling water	86 m^3		

gas can be used for various purposes, such as feedstock for chemical processes or as an energy carrier.

4.4.3 Acceptance Criteria for the Input Material

In principle, this process recycles all types of PVC waste. No differentiation between PVC formulae has to be made. There are no particular acceptance criteria. The waste is conditioned as follows to be acceptable for the slag bath gasifier:

- Intake and storage of the waste;
- Crushing and screening of the waste to the required particle size;
- Separation of iron and heavy non-ferrous metals from the waste by magnet or gravity sifter, respectively.

4.4.4 Environmental and Economic Performance

For the Linde process, a material and energy balance has been produced. If the input is some 3 tonnes PVC per hour, some 3,500-4,000 m^3 combustible gas and 700 m^3/h HCl (STP) is produced. No dioxins or furans are expected to be generated given the reducing atmosphere.

As for costs, ECVM estimates that considering all costs the total gate fee must be around €200 per tonne to break even for a 25,000 tpa plant. For pretreatment like grinding another €125 per tonne might be needed. However, it is clear that reliable estimates can only be given after experience with the pilot plant has been evaluated.

4.5 NKT Pyrolysis Process

4.5.1 Background and Current Status

In Denmark, the discussion on PVC waste is probably one of the most tense in the EU. Denmark was one of the first countries to have a covenant aiming to avoid the landfilling of PVC from building waste applications. Whereas for waste flows like pipes and window frames mechanical recycling is a good option, the situation is a bit more difficult for PVC isolation from cables. The NKT process was initially developed to deal with this waste flow. A first investigation on a laboratory scale started in 1993, followed by technical scale in 1995. Between 1998-1999 the Danish Environmental Protection Agency (EPA), the NKT holding, ECVM and the Norwegian company Norsk Hydro sponsored further investigation of the optimisation of the process for the treatment of mixed PVC building waste on a semi-technical scale. A pilot plant project financed by the Danish EPA and NKT was started in September 1998, involving the construction of a pretreatment plant for the treatment of about 1,000 tonnes/year mixed plastic waste and a reactor for the treatment of 200 tonnes/year of PVC waste. A further step might be upgrading to a 15,000 tonnes/year plant for mixed PVC waste depending on economic viability.

This project has been presented to the management committee of Vinyl 2010 (the organisation that manages the EU PVC industries recycling initiatives) with a request for financial support to scale-up to commercial size. A decision should be taken by the end of 2002, taking into account the total available quantities of PVC waste and the development of alternatives technologies in Denmark.

4.5.2 Description of the Process

PVC is transformed into various chemical products/ raw materials. In the pre-treatment step light plastics such as PE, PP, wood and the like are sorted out, as are sand, iron, steel, brass, copper and other metallic pollutants.

In a reactor at low pressure (0.2-0.3 MPa) and moderate temperature (375 °C) the PVC is chemically and thermally degraded. A particular feature of the process is that the chlorine in the PVC reacts in part with the fillers in PVC and is neutralised with the formation of $CaCl_2$. In similar vein, metal stabilisers in PVC are converted into the respective metal chlorides (lead, cadmium, zinc and/or barium). At current PVC waste compositions these chlorides consist of 60% lead which can be purified and re-used. The reaction in the end results in the following: solid, liquid and gaseous products.

From the gaseous product, HCl is recovered by absorption in water. The other gases (CO, propane, ethane) are incinerated and released. The liquid phase is separated into an organic condensate and an aqueous condensate. Solutions containing HCl can be reused in the downstream separation process. The solid phase is treated in a multistage extraction-filtration. Via control of pH, temperature and the amount of water added, heavy metals are separated from the coke. The chloride that is not separated or re-used earlier becomes available as calcium chloride from the evaporation step. Water is recycled in every extraction stage to minimise consumption.

In sum, the products of the process are:

- Calcium chloride product (< 1 ppm lead), which may be used as thaw salt or for other purposes
- Coke product (< 0.1 wt% of both lead and chlorine), which may be used as fuel in a cement kiln
- Metal concentrate (up to 60 wt% lead), which may be further purified and re-used
- Organic condensate, which may be used as fuel for the process.

4.5.3 Acceptance Criteria for the Input Material

A great variety of PVC materials have been tested in the process thus far: cable, cable trays, flooring material, window frames, artificial leather, packaging, pipes, flexible hoses, ring binders and roofing material. This indicates that the process is robust and can handle a broad range of PVC materials. In terms of chlorine content, there are no restrictions. With the completion of the pretreatment pilot plant, the contents of other plastics and metals may now be reduced significantly.

4.5.4 Environmental and Economic Performance

The process needs input of lime and water next to the PVC waste. No energy input is needed since the organic condensate provides for the energy needed in the process. Energy needed for pretreatment can be up to 25-35 kWh/tonne. Downstream separation of the coke products needs another 30-40 kWh/tonne. The process does not emit dioxins, metals or plasticisers. Due to internal recycling there are no aqueous waste streams. The reaction of lime with HCl forms some CO_2. The coke product provides a calorific value.

Cost data cannot be final at this stage and need further evaluation. It is expected that they will be around €250 per tonne for a 15,000 tonnes/year plant; such a plant would need an investment of some €10 million.

5 Dedicated Chemical Recycling for Specific Plastics

5.1 Introduction

Apart from MPW and PVC, several feedstock recycling processes have been developed for some specific plastics. In this respect, the most important plastics are:

- PET
- PUR
- Nylon.

The developments are discussed next.

5.2 PET

PET is an important plastic, which is currently gaining ground in the packaging field. It is the prime plastic used for drinking bottles. One of the key discussions is the question of whether returnable bottles are preferable to one-way bottles. The latter are often favoured by industry, since the logistics of a return system are perceived as problematic. However, policy makers often only accept one-way bottles under the condition that a good recycling scheme is set up. Mechanical recycling is one option, but recently also chemical recycling options have been proposed.

An important initiative for PET recycling is PETCORE, a consortium of some 15 major firms in the field of plastics, soft drinks and water, and packaging. Mechanical recycling is one of the major routes for recycling PET, but chemical recycling options have become operational as well. In such chemical recycling processes PET is broken down into its basic chemical building blocks, which are purified and then reassembled into new polymers. The advantage is that the product quality is not influenced by, e.g., the colour of the input, etc. Processes mentioned in the literature include the ones developed by the Eastman Chemical Company and DuPont de Nemours.

The Eastman process breaks down the PET down into basic components that can be separated from dyes, additives and other impurities. At this stage the pilot plant is still a rather small operation that is designed primarily to produce data rather than the product (97).

The DuPont patent (US 5866622 A, 1999) describes dissolving the polyester in molten dimethylterephthalate, methyl-*p*-toluate or dimethylisophthalate and separating the polyester from non-polyester components. The polyester can subsequently be used as a feedstock for methanolysis to form dimethylterephthalate (DMT) and alkylene glycol. The DMT can be subsequently hydrolysed to recover terephthalic acid.

Teijin, a Japanese company, has developed a similar process. They want to build a 30,000 t/y plant, operational by 2002.

5.3 PUR

Several companies have started to investigate whether chemical recycling of PUR is a viable option. In 1997, a PU glycolysis plant was started up at Philip Environmental Services in Detroit using technology patented by BASF. The plant is generating some 10 million pounds of glycolysate per year. The plant generates recycled polyols, which can be re-used in the production of PU.

Another initiative has been taken by the firm Regra (Germany), which has developed a process based on glycolysis. PU waste is chopped in a cutting mill and fed into an unpressurised reaction vessel. After the addition of glycols and additives, the temperature of the reactor is raised to 250 °C and the PU wastes are split through chemical and physical processes. The final product, after purification, is a liquid consisting of polyols and low molecular weight urethanes which, taken together, make the recovered polyol. These materials are capable of reacting with isocyanate and can be used to make new PU products (132).

ICI has also developed a PU recycling process for all MDI flexible PU foam slabstock. With their partner Vergier, they have established a UK£1 million pilot plant facility, aiming to move to commercial scale 5,000 tpa operation in 3 to 5 years after 1998. However, no information is available at this stage on plant at a commercial scale (131).

5.4 Nylon from Carpets

The last example of chemical recycling of a specific plastic is the recycling of nylon from carpets. The RECAM (recycling of carpet materials) project is one example. Floorings are collected and sorted by type, and particularly the nylon-based carpets are often recycled by some form of chemical recycling.

DSM and Allied Signal have developed carpet identification equipment that can identify specific plastics. Particularly carpet made of nylon 6 (or nylon 66) fibres are separated, cut into pieces, isolated, and converted by polycondensation into their monomeric components. The monomers can then be reused in the production of nylon. DSM and Allied Signal opened the world's first large scale carpet recycling plant in Augusta, Georgia, US in November 1999. This plant has a capacity of 90,000 tpa nylon 6 carpet waste (109).

Using this technology, the German company Polyamid 2000 Aktiengesellschaft is building a large-scale industrial plant close to the Polish border for obtaining nylon from used carpet (120,000 tpa carpet, which includes 24,000 tpa nylon). This implies a capital investment of DM310 million (some €155 million) (109).

DuPont has recently announced plans to build a demonstration plant in Maitland, Ontario, to show that the quality of the recycled product is equivalent to the virgin material (64). BASF converts post-consumer carpet into caprolactam in Ontario. Rhodia has several European plants for depolymerising nylon 6.

6 Other Treatment Options for Mixed Plastic Waste

6.1 Alternatives to Feedstock Recycling

Obviously, there are more traditional treatment options for plastic waste than feedstock recycling. In the next chapter I will compare the feedstock option with more traditional options, here some data are given on the latter. In this section, I will discuss the following technologies:

- Cement kilns;
- MSWIs (several flue gas cleaning options);
- Mechanical recycling;
- Landfill.

Furthermore, Solvay have developed a process called Vinyloop®, that is an intermediate between chemical and mechanical recycling and will also be discussed here.

6.2 The Vinyloop® PVC-Recovery Process

6.2.1 Background and Current Status

The Vinyloop® process was developed by Solvay as a response to a challenge from one of its customers, Ferrari Textiles Techniques (France), who produces architectural tarpaulin and canvas in PVC/polyester compound. This is a rather difficult formulation for recycling, since the PVC is mixed with a matrix. At the same time, Ferrari felt it was important that their material would be recyclable. The first Vinyloop® installation is now operational. It is a form of mechanical rather than feedstock or chemical recycling, since the PVC matrix is not changed in the process.

Around 1999, a 25 kg/day (about 1 tonne/year) experimental installation was set up. A 1,000 tpa installation followed, and by now Solvay is expected to have come close to completion of a full-scale plant of 17,000 plant.

6.2.2 Description of the Process

In principle, the process is quite simple. The input is cut and reduced in size. The key of the process is selective dissolution of the PVC and its additives in a special solvent. The pure PVC is recovered by means of precipitation and dried and is ready for a new life (**Figure 2**).

As indicated, this process is defined as mechanical recycling, since the PVC polymer is not broken down into its monomers. Yet, a main difference with classical mechanical recycling is that in the latter the full PVC formulation is kept intact. Here the components that make up the full formulation are separated. This results in the important advantage that the Vinyloop® process can deal with rather complicated formulations. According to Solvay the regenerated PVC is comparable in quality to the primary product.

6.2.3 Acceptance Criteria, Resource Needs and Emissions, and Costs

The input of the process is a separately collected PVC fraction. The quality has to be about similar as for PVC bound for mechanical recycling. The pilot scale tests

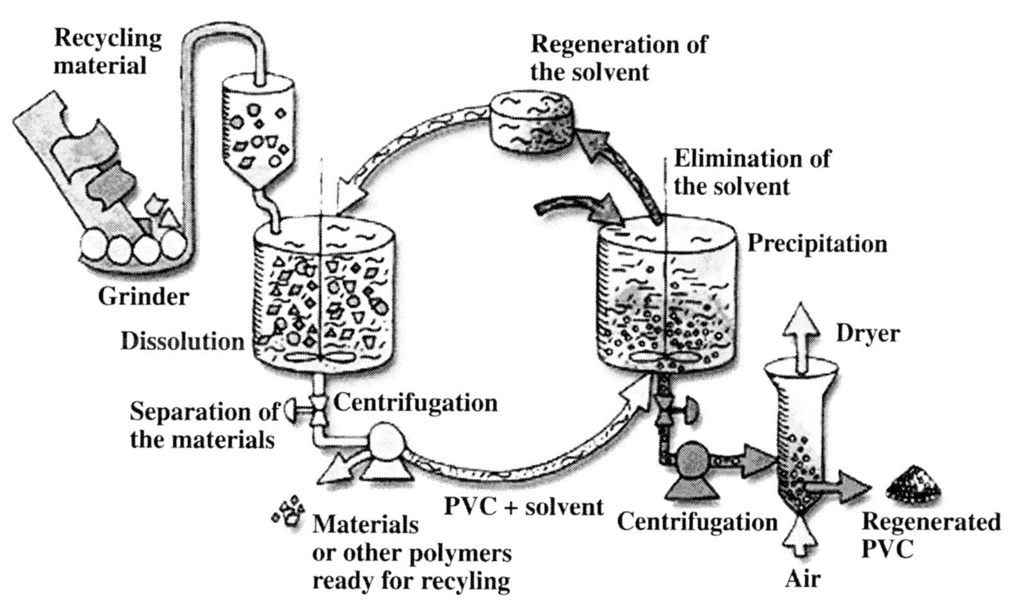

Figure 2

Schematic of Solvay Vinyloop® process

Reproduced with permission from Solvay

showed that the process can handle all PVC-compound materials tested so far: cables, pharmaceutical blister packs, floor coating, car dashboards, etc. Since the process is a closed system emissions are low. Details about the resource use (particularly the solvents, the crucial element in the process) are not publicly known. The gate fee is said to be about €350 per tonne.

6.3 Cement Kilns (Energy Recovery)

6.3.1 Background and Current Status

Cement production demands major amounts of fuel. Energy costs can be upwards of 25% of their turnover. Therefore, many cement kilns in the UK, Belgium, the Netherlands, Switzerland and other countries have started to use pretreated waste streams as a fuel. This not only saves fuel input, but indeed often allows a charge for the treatment of waste. In principle, cement kilns can deal with the following types of waste:

(1) liquid, high calorific fractions (as fuel)

(2) liquid, low calorific fractions

(3) sludges (as raw material and fuel)

(4) solid waste, including plastics (as raw material and fuel).

For waste types (1), (3) and (4) the use in cement kilns can be regarded as a recovery operation. For waste type (2), however, there is no real benefit of using the material in the cement making process. The kiln is merely being employed as a means for a (thermic) waste disposal operation.

Different cement kiln operators concentrate on different waste market segments. Some Belgian and UK cement kilns specialise in dealing with hazardous waste, and others concentrate on non-hazardous wastes like MPW. This has the disadvantage of a lower gate fee that can be charged, but the advantage of less public concern as compared to hazardous waste incineration. As for blast furnaces, the capacity of cement kilns to deal with MPW can be enormous. In Europe about 250 million tonnes of cement are produced annually, implying an energy need of some 800-1,000 billion MJ per annum (a.14). With the calorific value of MPW being some 30,000 MJ/tonne, this equals 30 million tonne of MPW. Even with 10% replacement of energy carrier by plastic waste, this would imply a capacity of 3 million tonnes per annum.

6.3.2 Description of the Process

Cement kilns produce a clinker by sintering alkaline raw materials such as lime ($CaCO_3$), clay (SiO_2 and Al_2O_3) and gypsum ($CaSO_4$) in a kiln at high temperature. Temperatures can be up to 1,450 °C in the solid fraction. The kiln is in fact a rotary kiln with a very long length (200 metres). In general, the solid materials flow in the opposite direction to the incineration gases. Due to the length of the kiln incineration gases have a long residence time at high temperatures (4 to 6 seconds at 1,800 °C and 15 to 20 seconds at 1,200 °C (a.15)). The oxygen content in the incineration process, however, is much lower than in normal waste processes.

There are two main processes for clinker production: a so-called wet process and a dry process. In the dry process the alkali raw materials are introduced in dry form into the kiln. In the wet process, these materials are introduced in the form of a slurry. Among other things, the type of process depends on the source of the kiln's raw materials. If a kiln extracts alkali raw materials from lakes then a choice for a wet process is logical, despite the disadvantage of its rather high energy use (5,000 MJ/tonne versus 3,600 MJ/tonne clinker), as in the dry process no water has to be evaporated.

The high temperatures ensure that inputs like MPW are effectively destroyed. Acidic substances such as HCl and SO_x are neutralised by the alkali raw materials, which act in fact as a caustic scrubber. Metals are bound in the clinker or in the fly ash. Fly ash is captured with an electrofilter and subsequently added to the clinker. In general, no other flue gas cleaning is applied.

6.3.3 Acceptance Criteria for the Input Material

Cement kilns can handle quite a lot of different materials. In most cases the input material should be chipped or shredded. Licenses often limit the PVC and chlorine input to 1%-2% chlorine in waste. It is said that chlorine also has a negative impact on the quality of the clinker if it is available in too high quantities. The content in clinker seems to be limited to some 0.1% at maximum, and hence the average chlorine content of all fuels used combined may have to be somewhat lower. This implies that waste with a high chlorine content has to be fed into the kiln simultaneously with wastes or fuel with a lower chlorine content.

6.3.4 Environmental and Economic Performance

For treatment of waste in cement kilns, several LCAs have been performed (a.16). Normally, a waste-independent mass balance model is applied. Given the specific composition of the particular waste, the model, based on a Belgian wet cement kiln, calculates the change in emissions to air and the components added to the clinker when waste instead of fuel is incinerated. For MPW, the energy content basically replaces coal or oil, and chlorine is essentially captured as chloride, along with other impurities, in the clinker. A point to be noted is the production of so-called particles of incomplete combustion. If the waste is clearly of a different quality compared to regular fuel, it is strongly suggested that trial burns and subsequent measurements should be carried out to prove that the emission of products of incomplete combustion (PIC) is not enhanced. This is often a concern of stakeholders living near the plant and can generate a lot of discussion.

As for costs, cement kilns have the advantage that they are primarily constructed for other purposes than waste incineration. Hence capital investment can be disregarded by cement kiln operators. The limited costs for pretreatment (shredding, etc.) will probably already be covered by saved expenses of the replaced fuel. In practice, cement kiln operators tend to concentrate on waste for which they can ask the highest price, while still being competitive with regular waste treatment options. In sum, the price for treatment in cement kilns will thus mainly depend on the availability and price of alternatives, and may range between a few Euro to €100 per tonne.

6.4 Municipal Solid Waste Incinerators (with Energy Recovery)

6.4.1 Background and Current Status

Municipal solid waste incinerators (MSWIs) are a robust treatment method for very different mixed waste types of different origin. The typical MSWI handles waste of a calorific value between 9 and 13 MJ/kg. They are the key technology for the treatment of integral household waste in countries such as Denmark, Sweden, the Netherlands and Germany. Some 7% of this integral household waste consists of plastics. Treatment of plastics waste as long as it is part of integral household waste is no problem in MSWIs. However, if plastics waste was collected separately and then submitted to a MSWI, problems could arise since pure plastics waste has a high calorific value (30 MJ/kg or more).

6.4.2 Description of the Process

In an MSWI the waste, after it is tipped into storage and has been made more homogeneous, is transferred to a grid-type kiln. The rolling grid is placed at a certain slope. This allows the waste to be transported with such a speed, that full incineration takes place. Finally, a slag remains which is treated in order to recover the ferrous and non-ferrous fraction. In some countries these slags are re-used, mainly in road construction. The flue gases pass through cleaning equipment such as an electrofilter, an acid scrubber, a caustic scrubber, an active carbon scrubber and a DeNOx (nitrogen oxide abatement) installation in order to comply with the demands of the EU incineration directive. In general, the energy is also recovered by electricity production or heat transfer (a.17). The flue gas cleaning leads to fly ash and flue gas cleaning residue, which has to be landfilled. A large fraction of the chlorine input into the MSWI ends up in the flue gas cleaning residue. In dry flue gas cleaning , the amount of flue gas cleaning residue can be as high as 66 kg per tonne of waste incinerated (e.g., a.11).

An alternative process has been developed for the neutralisation of flue gases with sodium bicarbonate. The amount of flue gas cleaning residue does not differ significantly in comparison with the regular flue gas cleaning. However, this residue can be treated at a separate plant recovering soda and salt, which in turn diminishes the amount of final waste greatly. This process is called the NEUTREC system and is operational at a Solvay plant in Rosignano in Italy. This system is applied in only a limited number of MSWIs.

6.4.3 Acceptance Criteria for the Input Material

MSWIs can accept virtually any mixed waste stream as long as it falls within its 'calorific window' of 9-13 MJ/kg (hence including material containing regular plastics and PVC content). Furthermore, the heavy metal content should not be excessive (since this can make the quality of the slag not suitable for re-use) and the material should be destructible at the rather low incineration temperature of MSWIs (850 °C). Some waste will not be efficiently destroyed.

6.4.4 Environmental and Economic Performance

Many LCAs have been performed for MSWIs (e.g., a.16). One can calculate the emissions and other environmental aspects of MSWIs for a given waste on the following basis:

- as a function of the composition of the waste: the component-related emissions to air, water and waste residues on the basis of the mass balances
- as a function of the calorific value of the waste: the process-related emissions to air and water and the energy recovery
- as a function of the ash content of the waste: the amount of slag and fly ash.

In general, emissions form no major problem. Slag and fly ash can be landfilled, though the metal content is a point that deserves attention in the long-term. The energy recovery is generally limited to some 20% electrical energy plus 10%-20% thermal energy as a maximum. The typical gate fees are between €100 and €150 per tonne, though it has to be noted that price dumping is possible.

6.5 Mechanical Recycling and Landfill

Finally, other relevant treatment options for plastics waste include landfill and mechanical recycling. Since these options (unlike Vinyloop and cement kiln incineration) are not even similar to feedstock recycling we discuss them here only very briefly. Mechanical recycling of plastics (be it PVC or other plastics), needs dedicated collection of the plastic waste in question. This is only possible for selected plastic flows (high volumes, recognisable products, products consisting mainly of one plastic). Landfill can accept plastic waste in any waste context (pure plastic type, MPW, mixed materials). I will only address the costs of these alternative technologies.

For landfill, costs vary highly across Europe. The main reason are differences in landfill quality (a point that will be quickly become less important now that the EU Landfill directive is becoming operational) and particularly landfill tax systems. Differences between some €2 per tonne in some EU countries (for inert waste) and up to €280 per tonne in, e.g., some places in Germany, may occur (though the latter value must be regarded as exceptionally high). Technically, even in countries with unfavourable circumstances like the Netherlands (high ground water level, soft soil), a price of some €50 per tonne is enough to realise a controlled landfill (including aftercare systems). The gate fee of mechanical recycling processes depends very much on the type of plastic and type of mechanical recycling and cannot be specified here.

7 Pros and Cons of the Different Treatment Routes

7.1 Introduction

Table 3 reviews the main aspects of the treatment technologies discussed in this section. The table summarises the type of material input, the maximum allowable chlorine content (often being one of the most crucial acceptance criteria), the gate fee, the status of the technology, and the products formed from chlorine, the organic fraction and metals. It should be noted that the gate fee reflects only the costs of final treatment (so-called 'hot box' and mechanical recycling processes). The costs for collection and pretreatment ('cold box' processes) still have to be added. Such costs differ greatly for each type of waste and each type of final treatment, and will be discussed together with the pros and cons of different treatment routes in this chapter.

As for environmental aspects, in the last years many LCAs have been performed that aimed to compare the different recycling and treatment routes for plastics packaging waste. They often include cost calculations for the whole waste management chain. Examples include:

- The 1994 CE study on disposal of municipal plastics waste (a.11) and a follow-up study that included the Texaco process (a.3)
- The 1999 Heyde and Kremer Frauenhofer LCA on plastics packaging waste (a.6)
- Various studies by TNO (feedstock recycling of plastics for the EU, the cold box study, and various LCAs for the Dutch packaging covenant and industrial clients)
- A forthcoming study by the Oko-Institut on the management of lightweight packaging.

Table 3 A review of technical aspects and gate fees for MPW treatment options (excluding collection/pretreatment)

Process	Input	Max. input Cl	Gate fee (€) Excl. collection /pretreatment (gate fee for a large scale plant in brackets)	Status	Products/fate 1. Organic fraction 2. Chlorine 3. Metals	Capacity	Future potential	Remarks
Texaco	MPW	5%	100 (50)	Pilot	1. Syngas 2. NH_4Cl-product 3. In vitrified slag	-	Uncertain*	
Polymer cracking	MPW	2%	200 (100-175)	Pilot	1. Liquid/gas 2. $CaCl_2$ (landfill) 3. In various residues	-	Uncertain*	5% chlorine possible for short periods
BASF	MPW	2.5%	250 (160)	Demo (closed)	1. Liquid/gas 2. HCl (product) 3. In residues	15 ktpa before 1996	-	On hold due to more economical competitors
Blast furnace	MPW	1.5%	Few-100?	Operational	1. Coal replacement 2. Cl (to water) 3. In iron or slag	162.5 ktpa in 1998	5 million tpa in the EU**	
Veba	MPW	2%	250	Operational (to be closed)	1. Gas/syncrude 2. HCl (product) 3. Hydrogenated residue	87 ktpa before 2000	-	Higher chlorine content possible for short periods
SVZ	MPW	2-5%	150?	Operational	1. Syngas/methanol 2. Cl to waste? 3. In landfill class 1 slag	110 ktpa in 1998		Higher chlorine levels are possible for short periods
MSWI	MSW ca	Not relevant	100-150	Operational	1. Energy (20-40%) 2/3. Cl and metals to waste	N/A	N/A	
Cement kilns	MPW	1-2%	Few-100?	Operational	1. Energy (100%) 2/3. Metals and Cl in cement	Some 100+ ktpa	3 million tpa in the EU**	
Mechanical recycling	Mono waste flow		Some 200+, much lower for cables	Operational	Recovered PE or PP	N/A	N/A	Highly product specific
Landfill	MSW ca	Not relevant	1-280	Operational	-	N/A	N/A	a.18

Note: MPW is Mixed Plastic Waste; MSW ca is municipal solid waste and comparable material
* Typical capacities considered are 50 ktpa to 200 tpa
** Theoretical potential if most blast furnaces and cement kilns in the EU start to replace regular fossil resources by MPW

7.2 Discussion of Environmental Effects

For outsiders, it sometimes seems that LCAs never come to a conclusion. It is true that by varying certain assumptions, sometimes opposing answers can be obtained. However, in essence the determinants of the ranking for MPW waste management technologies are pretty straightforward. A closer look at the energy potential of recovery and recycling indicates why.

It can be debated to what extent energy is a sufficient indicator for all potential environmental effects over the life cycle. However, energy use is in general the dominant contributor to emissions that score on acidification, global warming potential and other environmental effects. Furthermore, toxicity themes are still rather weakly developed in LCA. Hence, often energy use, the volume of final waste and land use give already a good first insight into the pros and cons of treatment routes.

Table 4 lists, for a number of the most common plastics, the total energy used for producing the primary resin. In general, this total is around 80-90 MJ/kg. It has to be noted, though, that this energy has to be split up into two main elements. The first element is the feedstock energy that in the end is embodied in the plastic itself. This is, not surprisingly, close to the calorific value of regular oil (some 40 MJ/kg). The second element, the remaining 40-50 MJ/kg, is the energy that is used in the production chain.

LCAs on treatment of plastics waste basically calculate the resultant of two parts:

(a) the environmental effects that are caused during collection and treatment; and

(b) a 'bonus' for the secondary products that are generated, since these do not have to be produced anymore via regular routes.

Table 4 Energy values related to some plastics (a.11)

Plastic type	Energy value
Total energy use for production virgin resin (feedstock + other)	
-HDPE	81 MJ/kg
-LDPE	89 MJ/kg
-PET	85 MJ/kg
MPW typical calorific value	38 MJ/kg

With this in mind, one can already say a lot about the maximum potential environmental bonuses per route:

(1) MSWIs use the calorific value of the MPW to produce heat and/or electricity.

(2) The Texaco, Polymer Cracking, VEBA and BASF processes all produce mainly liquid organics or gases that 'replace' primary oil- or gas-based resources. However, it has to be acknowledged that both BASF and VEBA have been closed down or will be closed down shortly, and that the other two processes have not yet been realised on a large scale.

(3) The SVZ process, blast furnaces and cement kiln all use the MPW as a replacement for coal. One could argue that in cement kilns other fuels are also replaced. However, coal appears to be the main energy source for cement kilns.

(4) Mechanical recycling uses MPW as a replacement for primary plastic resin.

For (1), MSWIs, the maximum bonus is limited by the calorific value of the plastics waste (about 40 MJ/kg). Furthermore, the energy recovery is relatively low due to technical limitations in comparison to normal power plants. Normally, at best some 20% electrical energy is recovered (or some 50%-70% calculated as primary energy).

In cases (2) and (3), the maximum bonus is limited by the calorific value of the plastics waste (40 MJ/kg) plus the energy needed to produce the replacement energy carrier (often just a few MJ/kg). However, particularly in the processes listed under (3) the efficiency of the energy recovery is 100%. (Formally at blast furnaces the plastics are used as reducing agents rather than energy carriers. However, the argument about efficiency still holds.) The feedstock recycling processes listed under (2) use some energy, but this is compensated by the quality gained in the output energy/feedstock carrier.

In case (4), potentially the full 80+ MJ/kg which encompasses feedstock and the fairly high energy requirement for producing plastics can be recovered. However, here various complications arise:

(a) The collection, separation and cleaning is often more energy-intensive.

(b) The fraction that is really used as secondary plastics is often far below 100%, the rest has to be sent to one of the other options after separation.

(c) The secondary plastics often do not have the same quality as primary plastics, or cannot be used in high-quality products, so for the fraction that is recycled less than 100% of the potential 80 MJ/kg can be attributed.

Figure 3 puts this reasoning in perspective. For each type of option, the left bar in the chart indicates the maximum potential energy benefit (i.e., feedstock plus production energy of the product replaced). The right bar gives the achievable benefit. For example, the recovery efficiency of an MSWI is around 50%, and there is a moderate efficiency of the cleaning/separation process for mechanical recycling of say 50% with a moderate replacement factor of 70% compared to virgin plastics (i.e., secondary plastic is generally of lower quality compared to virgin). In this Figure, the energy input side (for collection and treatment) is not included.

And indeed, one sees this picture in virtually all LCAs. Direct incineration in MSWIs in virtually all cases scores worse than feedstock recycling or 100% use as energy carrier/reducing agent. The relatively limited additional effort for the post-collection sorting that is needed for feedstock recycling or use as fuel/reduction agent is by far outweighed by the much better energy recovery. Feedstock recycling processes and use as energy or reducing agent in general score quite similarly. Mechanical recycling is only a truly better option if one is able to use technologies that lead to high-quality secondary material. Otherwise due to more complicated collection and upgrading, there is only a low effective replacement of primary by secondary material and a low fraction that eventually is used as secondary plastic.

7.3 Discussion of Economic Aspects

Table 5 is based on the TNO study 'Chemical recycling of plastics waste' (a.1). It gives a tentative comparison of the costs of the different treatment routes.

Many initiatives in the field of feedstock recycling have been taken, it has proven difficult to reach a stable position in the waste management market. The VEBA, BASF and SVZ processes are the only ones that are, or have been, available in practice – a telling fact, since these are all German initiatives for which the stringent German Packaging decree via its very high recycling targets created a market. However, even in these rather favourable circumstances two initiatives ceased to exist. The BASF initiative has been put on hold, and the VEBA process. The most likely reason is that they depend(ed) on contracts with DSD, which were not extended since DSD found more cost-effective

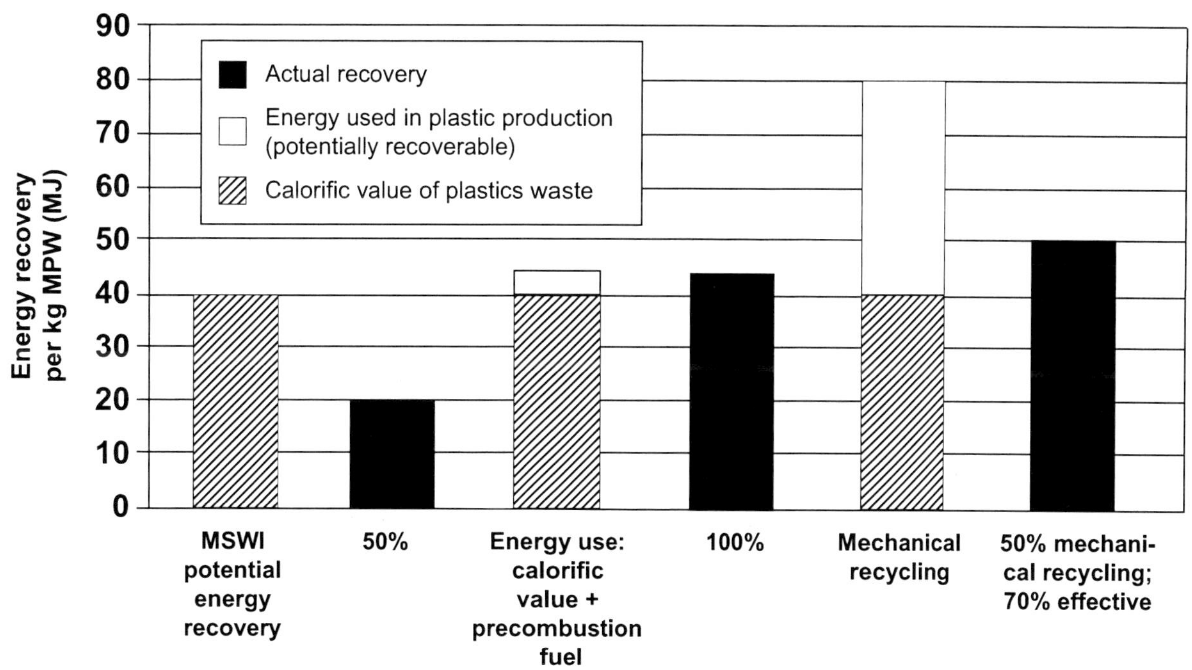

Figure 3

Potential maximum energy recovery versus practical achievable energy recovery of various plastics waste management routes

treatment options (i.e., SVZ and blast furnaces). Furthermore, the initiatives for the Texaco process and the polymer cracking process have also, after several years, not yet lead to investments in a full-scale plant. Rather, there are indications that potential waste suppliers initially interested in such a technology finally stepped back, since they found more cost-effective outlets for their MPW.

This whole picture suggests that chemical recycling is financially still a rather uncertain business, a view that is indeed reinforced by cost calculations. The basic point probably is that purpose-built recovery installations will always have trouble in competing with technologies built for another purpose, but which happen to be able to recycle or recover MPW – which is the case with blast furnaces and cement kilns. The latter have the advantage that capital investment does not need to be allocated to the MPW.

8 Overall Conclusions

For treatment of plastics packaging waste, many technologies are available. They basically can be divided into the following classes:

(1) Integrated collection with household waste and incineration in an MSWI with energy recovery

(2) Separation of plastics with a high-calorific fraction from household waste and use as energy carrier in, e.g., cement kilns

(3) Separation of plastics such as from household waste and use as reducing agent in blast furnaces or for feedstock recycling

(4) Separate collection of plastics, sorting, cleaning and mechanical recycling.

From the comparison it follows that route (1), direct incineration in MSWIs, is generally the least favourable option in environmental terms. Hence, there is no doubt that the targets in the EU Packaging directive, which aim to steer away from incinerating plastics packaging in MSWIs, have a sound rationale.

As for routes (2) and (3), the conclusion in this paper, as one can find back in many LCAs, is that the difference in environmental terms is not big. After all, both routes make use of the energy content (or carbon content) of the plastics, with a (close to) 100% efficiency. This is a striking point, since route (2) is labelled as 'incineration with energy recovery' whereas route (3) is labelled as 'recovery', and hence in legal terms an option that scores

Table 5 Tentative cost comparison of treatment of plastics waste (in €/ton)

	Landfill	MSWI	Cement kiln	Blast furnaces	Texaco, VEBA, SVZ, Polymer cracking
Acceptance criteria:					
Max. PVC input	Not relevant	Not relevant	1%-2%	1%-2%	10%
Typical waste accepted	MSW	MSW	MPW, refuse derived fuel	MPW	MPW
Collection and pretreatment					
Integrated collection	160	160	160	160	160
Pretreatment/separation			25		
Agglomeration, separation				150*	150*
Logistics (i.e., transport)	40	40	40	40	40
Final treatment	50	125	50	50	150*
Total costs	250	325	275	400	500
* Can vary considerably depending on capacity or if no agglomeration is needed					

better. This is particularly relevant since dedicated feedstock recycling plants apparently can hardly compete with plants like cement kilns or blast furnaces, since in the latter case most or all of the capital investment is done for another purpose. Feedstock recycling plants in principle could produce higher valued products, but in practice all initiatives have lost to the competition up until now. On this point, one could question whether the distinction that the EU Packaging directive and other directives make between feedstock recycling (a recovery operation and the preferred option) and the above options with 100% energy recovery makes sense.

Option (4), mechanical recycling, is favourable under the condition that via advanced separation and upgrading technologies high-quality recycling can be reached. But if that is not the case, this route might not be convincingly better than route (2) or (3). It is clear that this option has the highest potential, therefore the protection via the recycling quota in the Packaging directive is in principle justified. However, for that fraction of the plastics waste for which high-quality mechanical recycling is not likely to be feasible, one should not pursue mechanical recycling at all costs. To put it bluntly: if (mechanical) recycling targets in regulations only lead to 'downcycling' like the use of secondary plastics in park benches, fancy ballpoint pens and roadside bollards, one has created a system that does not make sense from an environmental viewpoint. Despite the legal classification, both for environmental and cost reasons such plastics can be much better incinerated with 100% energy recovery.

Literature

a.1 A. Tukker, H. De Groot, L. Simons and S. Wiegersma, *Chemical Recycling of Plastics Waste (PVC and other Resins)*, 1999, TNO-STB Report 99-55, Delft, The Netherlands (www.europa.eu.int).

a.2 DKR/DSD, *Mit Altkunststoffen Ressourcen schonen*, DKR, Germany, 1999.

a.3 R. Weissman, Recycling of Mixed Plastic Waste by the Texaco Gasification Process in *Chemical Aspects of Plastics Recycling*, Eds., W. Hoyle and D.R. Karsa, Royal Society of Chemistry, Cambridge, UK, 1997.

a.4 H. Croezen and H. Sas, *Evaluation of the Texaco Gasification Process for Treatment of Mixed Household Waste*. Final report of phase 1 & 2, CE, Delft, the Netherlands, 1997.

a.5 J.H. Brophy, S. Hardmann and D.C. Wilson, Polymer Cracking for Feedstock Recycling of Mixed Plastic Wastes in *Chemical Aspects of Plastics Recycling*, Eds., W. Hoyle and D.R. Karsa. Royal Society of Chemistry, Cambridge, UK, 1997.

a.6 M. Heyde and S. Kremer, *LCA Packaging Plastics Waste*, LCA Documents, 1999, Vol. 2, No. 5, EcoMed, Landsberg, Germany.

a.7 J. Janz and W. Weiss, Injection of Waste Plastics into the Blast Furnace of Stahlwerke Bremen. *La Revue de Métallurgie-CIT*, 1996, 1219-1226.

a.8 H.G. Rigo, A. Chandler and W.S. Lainer, *The Relationship Between Chlorine in Waste Streams and Dioxin Emissions from Waste Combustor Stacks*, The American Society of Mechanical Engineers, CRTD No. 36, New York, USA, 1995.

a.9 L.P.M. Rijpkema and J.A. Zeevalking, *Specific Processing Costs of Waste Materials in a Municipal Solid Waste Incinerator*, Report TNO-MEP R96/248, TNO-MEP, Apeldoorn, the Netherlands, 1996.

a.10 P. Costner, *Chlorine and Dioxin: The Burning Question*, Greenpeace USA, Washington DC, USA, 1997.

a.11 H.J.W. Sas (Ed.), Verwijdering van Huishoudelijk Kunststofafval: Analyse van Milieu-effecten en Kosten *[Disposal of Municipal Plastic Waste: Analysis of Environmental Effects and Costs]*, CE, Delft, the Netherlands, 1994.

a.12 HCL, APME, *Veba Report Success in Depolymerising WEEE*, Harriman Chemconsult Ltd, February 1998, 15.

a.13 H. Pohle, *PVC und Umwelt. Eine Bestandsaufnahme*, Springer Verlag, Berlin, Germany, 1997, 120.

a.14 A. Caluori, F. Mark, M. Moser and A. Prisse, *Plastics Waste – An Alternative Fuel Source for Cement Production*, BCU, Switzerland, 1997.

a.15 Ciments d'Obourg, *Information about Mass Balances and Emissions from Cement Kiln 8 and 9 of Ciments d'Obourg, Belgium*, 1995.

a.16 A. Tukker, (Ed.), Milieu-effectrapport Meerjarenplan Gevaarlijke Afvalstoffen 1997-2007 *[Environmental Impact Assessment*

Multi-year Hazardous Waste Management Plan 1997-2007], Ministry of Housing, Physical Planning and Environment and the Inter-Provincial Union, the Hague, the Netherlands, 1996.

a.17 AOO, *Milieu-effectrapport Tienjarenprogramma Afvalstoffen 1995-2005, [Environmental Impact Assessment of the Ten-Year Waste Programme 1995-2005]*, AOO, Utrecht, the Netherlands, 1995.

a.18 Argus, *The Behaviour of PVC in Landfill*, EU DG ENV, 2000 (www.europa.eu.int).

Abbreviations and Acronyms

APME	Association of Plastic Manufacturers in Europe
CE	Dutch Centrum voor Energiebesparing en Schone Technologie
CFB	circulating fluid bed
DG III	EU Directorate General III
DKR	Deutsche Kunststof Recycling AG
DSD	Duales System Deutschland
E&E	electrical and electronic
ECVM	European Council of Vinyl Manufacturers
EDC	ethylene dichloride
EPA	Environmental Protection Agency
IVV	Frauenhofer Institut Verfahrenstechnik und Verpackung
KAB	Kohleöl Anlage Bottrop
LCA	life-cycle-assessment
MPW	mixed plastic waste
MPWI	mixed plastic waste incineration
MSW	municipal solid waste
MSWI	municipal solid waste incineration
PCB	polychlorinated biphenyl
PE	polyethylene
PET	polyethylene terephthalate
PIC	products of incomplete combustion
PP	polypropylene
PUR	polyurethane
PVC	polyvinyl chloride
RECAM	recycling of carpet materials
SVZ	Sekundärrohstoff Verwertungs Zentrum
TEQ	toxic equivalent
VCC	Veba Combi Cracking

Abstracts from the Polymer Library Database

Item 1

Polyurethanes Expo 2001. Creating Opportunity through Innovation. Proceedings of a conference held Columbus, Oh., 30th. Sept. - 3rd. Oct. 2001.
Arlington, Va., Alliance for the Polyurethanes Industry, 2001, Paper 1, p.3-9.

COATINGS PREPARED FROM POLYURETHANE SOFT FOAM RECYCLING POLYOLS

Langenstrassen R; Huth H; Pohl M; Schmidt K-H; Behrendt G; Ivanyi S I; Goering H
Wildau,Technische Fachhochschule; Sofia,University of Chemical Technology & Metallurgy; Bundesanstalt fuer Materialforschung & -Pruefung
(American Plastics Council; Alliance for the Polyurethanes Industry)

Recycled polyols are derived from polyurethane soft foam waste by the combination of glycolysis and aminolysis for the production of polyurethane coatings. The chemical recycling process produces homogeneous polyols of low glycol content, the hydroxyl number of which can be adjusted to the properties of the coatings required. Coatings are produced by simple mixing of the recycled polyols plus additives in low concentrations with a di- or poly- isocyanate. Hardness and elasticity of the coatings can be adjusted by variation of the isocyanate index. The combined aminolysis and glycolysis procedure is described. 5 refs.

BULGARIA; EASTERN EUROPE; EUROPEAN COMMUNITY; EUROPEAN UNION; GERMANY; USA; WESTERN EUROPE

Accession no.846271

Item 2

Macromolecular Materials and Engineering
286, No.11, 30th Nov.2001, p.695-704

UV CURABILITY AND MECHANICAL PROPERTIES OF NOVEL BINDER SYSTEMS DERIVED FROM POLY(ETHYLENE TEREPHTHALATE)(PET) WASTE FOR SOLVENTLESS MAGNETIC TAPE MANUFACTURING. I. ACRYLATED OLIGOESTERS

Farahat M S; Nikles D E
Alabama,University

PETP waste obtained from beverage bottles was depolymerised by glycolysis using diethylene glycol(DEG) in the presence of manganese acetate as a transesterification catalyst. Glycolysis was conducted at two different molar ratios of PETP/DEG(1:2.15 and 1:1.03) to give oligoester polyols of different molec.wt. Modification of the products by acrylation gave curable acrylated oligoesters. The curability of these oligomers under UV irradiation in the presence of 2-benzyl-2-dimethylamino-1-(4-morpholinophenyl)-1-butanone as a photoinitiator was examined. The mechanical properties were investigated. The measured tensile properties were in the range 4.62 to 45 MPa for maximum TS and 0.074 to 2.0 GPa for Young's modulus. 42 refs.

USA

Accession no.845957

Item 3

ACS Polymeric Materials Science and Engineering Fall Meeting.Volume 85.
Chicago, IL, 26th-30th August 2001, p.506-7.012

CURING BEHAVIOR OF EPOXY RESIN WITH AMINOLYSIS PRODUCTS OF WASTE POLYURETHANES

Lee D S; Hyun S W
Chonbuk,National University
(ACS,Div.of Polymeric Materials Science & Engng.)

Rigid polyurethane foams were prepared at room temperature using commercial polyols and polymeric 4,4'-diphenyl methane diisocyanate, and used to study their recycling by aminolysis. The reaction products obtained by treatment with diethylene triamine at 180 C were evaluated as hardeners for epoxy resins. The exothermic heats of curing were determined over the temperature range 60-80 C by differential scanning calorimetry. A reaction order of 2.2-2.4 was obtained. 8 refs.

KOREA

Accession no.845621

Item 4

Polymer International
51, No.2, Feb.2002, p.183-9

MECHANICAL CHARACTERISTICS OF MODIFIED UNSATURATED POLYESTER RESINS DERIVED FROM POLY(ETHYLENE TEREPHTHALATE) WASTE

Farahat M S
Alabama,University; Egyptian Petroleum Research Institute

The effect of incorporating p-hydroxybenzoic acid (I) into the structures of various unsaturated polyesters synthesised from polyethylene terephthalate (PET) waste depolymerised by glycolysis at three different diethylene glycol (DEG) ratios with Mn acetate as transesterification catalyst, was studied. Copolyesters of PET modified using various I mole ratios showed excellent mechanical and chemical properties because of their liquid crystalline behaviour. The oligoesters obtained from the twelve modified unsaturated polyesters (MUP) were reacted with I and maleic anhydride, with variation of the I ratio with a view to determining the effect on mechanical characteristics of the MUP. The results indicated that increasing the I molar ratio within the concentration range studied resulted in a pronounced increase in the mechanical properties of the MUP, mainly an increase in the values of the maximum compressive strength and the Young's modulus. 51 refs.

EGYPT; USA

Accession no.845464

Item 5

Warmer Bulletin
No.83, March 2002, p.20-1

USING WASTE PLASTICS AS A SUBSTITUTE FOR COAL

Coal used in power stations has the potential to be partly replaced by fuels derived from pre-treated plastics and paper waste, reducing both dependency on fossil fuels and reliance on landfill. APME reports on a project in the Netherlands which it co-sponsored to develop a substitute fuel from plastics. The environmental assessment of the project compared the environmental impacts of coal substitution with other plastics recovery methods, including gasification in feedstock recycling and energy recovery from plastics waste in cement kilns. The study also compared coal substitution with the generation of power from burning biomass.

APME

EUROPEAN COMMUNITY; EUROPEAN UNION; NETHERLANDS; WESTERN EUROPE

Accession no.845316

Item 6

Vinyltec 2001. Pushing Profitability. Conference proceedings.
Iselin, N.J., 11th-12th Sept.2001, p.79-87

VINYLOOP. A NEW PROCESS TO REGENERATE PVC COMPOUNDS FROM COMPOSITE RESIDUES. A NEW PRODUCT. THE PRECIPITATED PVC COMPOUND

Crucifix P
Solvay SA
(SPE,Vinyl Div.; SPE,Palisades Section)

In November 1997, Ferrari (France) invited Solvay to help them to solve a problem of recycling PVC coated textile for light mobile structures, especially ones used for short-term applications (advertising). The Vinyloop is a physical recycling process making use of an organic solvent to separate the PVC compound from the other materials in a PVC composite. It is a closed loop process; the solvent is completely recycled. Aspects outlined include history and a description of the process, precipitated PVC compound evaluation and financial data

BELGIUM; EUROPEAN COMMUNITY; EUROPEAN UNION; WESTERN EUROPE

Accession no.845034

Item 7

Journal of Applied Polymer Science
82, No. 1, 3rd October 2001, p.99-107

ALKALINE DEPOLYMERISATION OF POLY(TRIMETHYLENE TEREPHTHALATE)

Joon Ho Kim; Joon Jung Lee; Ji Young Yoon; Won Seok Lyoo; Kotek R
Yeungnam,University; North Carolina,State University

The effects of reaction media, composition and temperature on the rate of alkaline depolymerisation of poly(trimethylene terephthalate) (PTT) were investigated to assess its value as a chemical recycling process. The alkaline depolymerisation of PTT was carried out at 160-190C in ethylene glycol (EG), diethylene glycol (DEG), triethylene glycol (TEG), ethylene glycol monobutyl ether (EGMBE), diethylene glycol monoethyl ether (DEGMEE) respectively and a mixture of these solvents. The reaction quantitatively converts PTT to disodium terephthalate and 1,3-propanediol. It appears to occur in two stages: (1) an induction period, (2) PTT weight loss occurs linearly with increasing length of exposure to sodium hydroxide. There are three factors which increase the rate of depolymerisation: (1) higher reaction temperature and longer reaction time, (2) a significant increase occurs by adding ethereal solvents in the order of EG less than DEG less than TEG less than EGMBE less than DEGMEE, and (3) the decrease in the difference between the solubility parameters of PTT and the reaction solvents, which decrease in the order EG greater than DEG greater than TEG greater than EGMBE greater than DEGMEE. The investigation confirms that PTT waste can be successfully converted into useful products. 31 refs.

KOREA; USA

Accession no.844689

Item 8

European Plastics News
29, No.2, Feb.2002, p.24

SUSTAINED PERFORMANCE

Vink D

The German packaging ordinance introduced in 1991 set out to reduce the amount of packaging in circulation and to recycle the waste into new packaging or raw materials. The ordinance was extended in 1998 to allow energy recovery and now requires 60% of sales packaging to be recycled. However, the German plastics industry is concerned about the introduction of more specific proposals, such as setting quotas for specific materials, and not giving the industry the freedom to choose the most appropriate methods of collection and recycling.

EUROPEAN COMMUNITY; EUROPEAN UNION; GERMANY; WESTERN EUROPE

Accession no.842381

Item 9
Bristol, UK,Environment Agency, 2001, pp.53, 30cm, 1744
PLASTICS IN THE ENVIRONMENT
(UK,Environment Agency)

This is part of the environmental issues series of the UK Environment Agency. The report provides an overview of plastics looking at manufacture, uses and disposal. The aim of the report is to make recommendations on ways to ensure that society's use of plastics is more sustainable in the future.

EUROPEAN COMMUNITY; EUROPEAN UNION; UK; WESTERN EUROPE

Accession no.841684

Item 10
Polymer Degradation and Stability
75, No.1, 2002, p.185-91
STUDY ON METHANOLYTIC DEPOLYMERIZATION OF PET WITH SUPERCRITICAL METHANOL FOR CHEMICAL RECYCLING
Yong Yang; Yijun Lu; Hongwei Xiang; Yuanyuan Xu; Yongwang Li
Chinese Academy of Sciences

Polyethylene terephthalate (PET) was subjected to methanolytic depolymerisation with supercritical methanol in a stirred stainless steel autoclave at temperatures of 523-543 deg.C, pressure 8.5-14.0 MPa, and a 3-8 methanol to PET weight ratio. The solid products obtained, consisting mainly of dimethyl terephthalate and small amounts of methyl-(2-hydroxyethyl) terephthalate, bis(hydroxyethyl) terephthalate, dimers, and oligomers, were analysed by high performance liquid chromatography (HPLC), and the liquid products, mainly ethylene glycol and methanol were analysed by gas chromatography (GC). The temperature, weight ratio of methanol to PET, and the reaction time had a very marked effect on dimethyl terephthalate yield and the degree of PET depolymerisation, but the effect of pressure was insignificant above the methanol critical point. The optimum PET depolymerisation conditions were: temperature 533-543 K, pressure 9.0-11.0 MPa, and a methanol to PET weight ratio of 6-8. The depolymerisation of several PET wastes from the Chinese market was studied under the optimum conditions. 28 refs.

CHINA

Accession no.841527

Item 11
Asia Pacific Coatings Journal
14, No.6, Dec. 2001, p.557-63
DECOMPOSITION OF FIBER REINFORCED PLASTICS USING FLUID AT HIGH TEMPERATURE AND PRESSURE
Sugeta T; Nagaoka; Otake K; Sako T
Japan,National Institute of Advanced Industrial Science & Technology; Kumamoto,Industrial Research Institute; Shizuoka,University

An investigation is reported of the decomposition of fibre-reinforced plastics, being refractory waste, using a supercritical water and alkali solution with alcohol at high temperature and pressure. Fibre-reinforced unsaturated polyester was treated by supercritical water at 380 degrees C and most of the matrix was decomposed during 5 minutes reaction time. The main products were carbon dioxide and carbon monoxide in gas phase, and styrene derivatives and phthalic acid in liquid phase. After the treatment with supercritical water for 5 minutes, no significant change in the fibre recovered was detected using scanning electron microscopy or infrared spectroscopy. On the other hand, phenolic resin used as a matrix of CFRP was not decomposed using only supercritical water, but was promoted by supercritical water with alkali. Futhermore, with used of alcohol-alkali aqueous solution at a high temperature, phenolic resin was found to be mostly broken down to soluble products. 15 refs.

JAPAN

Accession no.840528

Item 12
Macromolecular Materials and Engineering
286, No.10, 25th Oct.2001, p.640-7
POLY(ETHYLENE TEREPHTHALATE) RECYCLING AND RECOVERY OF PURE TEREPHTHALIC ACID. KINETICS OF A PHASE TRANSFER CATALYZED ALKALINE HYDROLYSIS
Kosmidis V; Achilias D S; Karayannidis G P
Thessaloniki,Aristotle University

PETP flakes produced from used soft drinks bottles were subjected to alkaline hydrolysis in aqueous sodium hydroxide. A phase transfer catalyst (trioctylmethylammonium bromide) was used to enable the depolymerisation reaction to take place at room temperature and under mild conditions. The effects of temperature, alkali concentration, PETP particle size, PETP concentration and catalyst to PETP ratio on the reaction kinetics were studied. The disodium terephthalate produced was treated with sulphuric to give terephthalic acid of high purity. A simple theoretical model was developed to describe the hydrolysis rate. 17 refs.

EUROPEAN COMMUNITY; EUROPEAN UNION; GREECE; WESTERN EUROPE

Accession no.839665

Item 13
Macromolecular Rapid Communications
22, No.16, 7th Nov. 2001, p.1325-9
ORGANIC SOLVENT-FREE ENZYMATIC TRANSFORMATION OF

POLYCAPROLACTONE INTO REPOLYMERIZABLE OLIGOMERS IN SUPERCRITICAL CARBON DIOXIDE
Matsumura S; Ebata H; Kondo R; Toshima K
Yokohama,Keio University

Details are given of the enzymatic transformation of polycaprolactone into repolymerisable oligomers in supercritical carbon dioxide. The object was to establish a sustainable chemical recycling system for polycaprolactone. 14 refs.

JAPAN

Accession no.838167

Item 14
Macromolecular Materials and Engineering
286, No.9, 28th Sept.2001, p.513-5
SYNTHESIS OF URETHANE OIL VARNISHES FROM WASTE POLY(ETHYLENE TEREPHTHALATE)
Mecit O; Akar A
Istanbul,Technical University

Waste PETP was depolymerised by glycolysis to give hydroxyl-terminated oligomers(DPET), which were used in the synthesis of urethane oils. The effect of depolymerisation temps., the type of glycol and the amount of catalyst on the yield and composition of the depolymerisation products was studied. The physical properties of the urethane oils were compared with those of a commercially-available product. The reaction of DPET with isocyanates produced random linkage between different molecules with or without terephthaloyl groups. 15 refs.

TURKEY

Accession no.836241

Item 15
Polymer Engineering and Science
41, No.9, Sept.2001, p.1457-70
REVIEW: RECYCLING OF NYLON FROM CARPET WASTE
Mihut C; Captain D K; Gadala-Maria F; Amiridis M D
South Carolina,University

The problem of carpet recycling is considered and the different methods being proposed or commercially utilised are discussed. The main component of the carpet waste is fibres of nylon-6 and nylon-66. The review of the literature includes a limited amount of journal publications, which focus primarily on fundamental aspects, and a large number of patents, which describe the available technologies. The most promising recycling techniques (depolymerisation, extraction, melt blending and mechanical separation) are described. 48 refs.

USA

Accession no.836146

Item 16
Industrial & Engineering Chemistry Research
40, No.22, 31st Oct. 2001, p.4749-56
MICROWAVE-INDUCED PYROLYSIS OF PLASTIC WASTES
Ludlow-Palafox C; Chase H A
Cambridge,University

The performance of a novel microwave-induced pyrolysis process was evaluated by studying the degradation of HDPE and aluminium/polymer laminates in a semibatch bench-scale apparatus. The relationship between temperature, residence time of the pyrolytic products in the reactor, and the chemical composition of the hydrocarbon fraction produced was investigated. 28 refs.

EUROPEAN COMMUNITY; EUROPEAN UNION; UK; WESTERN EUROPE

Accession no.835576

Item 17
Polymer Recycling
6, No.1, 2001, p.49-56
WASTE PET - DERIVED SUBSTANCES AS CROSSLINKING AGENTS FOR EPOXY RESINS
Fabrycy E; Spychaj T; Pilawka R
Szczecin,Technical University

The products of the chemical degradation of PETP with triethylene tetramine and triethaneolamine can be used as epoxy resin hardeners, it is demonstrated. Products of PETP aminolysis with triethylene tetramine and aminoglycolysis with triethanolamine, were characterised using NMR and rheometric measurements. Characteristics of the crosslinking process for the system: epoxy resin/PETP/amine degradation product, and epoxy resin/TETA for comparison were investigated by DSC. Three classes of liquid epoxy resins based on bisphenol A, bisphenol F and epoxy novolak resins were used in the experiments. 16 refs.

EASTERN EUROPE; POLAND

Accession no.834286

Item 18
Polymer Recycling
6, No.1, 2001, p.43-8
INFLUENCE OF OXYGEN ON THE STEAM GASIFICATION OF PVC
Van Kasteren J M N; Slapak M J P
Eindhoven,University

The recycling of PVC by hydrothermal techniques is described, in which PVC is thermally converted in a steam atmosphere into hydrogen chloride, hydrogen, carbon dioxide, carbon monoxide and some gaseous and liquid hydrocarbons. Whilst gasification with only steam is an endothermic reaction, partial combustion of PVC by the addition of small amounts of air, enables autothermic operation of the process to take place. This work deals with the effects of the addition of air on the gasification products, and compares an endothermic operation with autothermic and exothermic operation. 11 refs.

EUROPEAN COMMUNITY; EUROPEAN UNION; NETHERLANDS; WESTERN EUROPE

Accession no.834285

Item 19

Polymer Recycling

6, No.1, 2001, p.35-41

ROLE OF SOME PROCESS VARIABLES IN THE OPERATION OF FLUIDIZED BED PYROLYSERS OF PLASTICS WASTES

Arena U; Mastellone M L

Naples,Second University

The results of research into the fluidised bed pyrolysis of plastic wastes are reported, with reference to determining the optimum process conditions for the process with respect to the reactor behaviour. The study investigates the effects of process variables such as bed temperature, polymer feed rate, bed hold-up, fluidising velocity, and size of inert material. Findings illustrate the importance of the knowledge of the hydrodynamics of the fluidised bed and of the interactions between bed and polymer particles in the design and operation of the reactor. 15 refs.

EUROPEAN COMMUNITY; EUROPEAN UNION; ITALY; WESTERN EUROPE

Accession no.834284

Item 20

Nippon Gomu Kyokaishi

74, No.5, May 2001, p.173-8

Japanese

CHEMICAL RECYCLING PROCESS FOR WASTE PLASTICS USING SUPER-CRITICAL WATER

Fukuzato R

The reaction-catalysing properties of super-critical fluids are described, and some examples are demonstrated of the chemical recycling of waste plastics. 16 refs. Articles from this journal can be requested for translation by subscribers to the Rapra produced International Polymer Science and Technology.

Accession no.834123

Item 21

Polymer Degradation and Stability

74, No.1 2001, p.25-32

HYDROGENOLYTIC DEGRADATION OF THERMOSETS

Braun D; von Gentzkow W; Rudolf A P

Deutsches Kunststoff Institut

The hydrogenolytic degradation of thermosets by partially hydrogenated aromatics such as tetraline and 9,10-dihydroanthracene was investigated as a means of recycling crosslinked polymers. Glass fibre-reinforced epoxy resin covered with copper foil was cleaved by reaction at 340 C for 2 h, giving more than 99 wt% soluble products. Reaction products including phenol, p-isopropylphenol and phthalic anhydride were quantitatively analysed by gas chromatography. It was not necessary to grind the resin, and the fibre reinforcement and copper foils were recovered. The hydrogenolysis also satisfactorily liquefied phenolic resins, melamine resins and crosslinked unsaturated polyesters. 6 refs.

EUROPEAN COMMUNITY; EUROPEAN UNION; GERMANY; WESTERN EUROPE

Accession no.832483

Item 22

Chemical and Engineering News

79, No.38, 17th Sept.2001, p.22

LEGISLATION PRODS FOAM RECYCLING

Reisch M

Last September, the European Parliament passed a directive calling for member countries to adopt legislation requiring that 80% of an automobile by weight be recycled by 2006. The number rises to 85% by 2015. Dow is exploring two processes for recycling the PU foam used in car seats, in headrests, behind dashboards and under carpets. One is a mechanical pulverisation process, the other is a solvolysis process that dissolves the foam in dicarboxylic acid to produce a "recyclate". The first process, developed by Mobius Technologies, uses specially designed equipment to pulverise PU foam into a fine powder at room temperature. The recycled powder can then be added to virgin foam to represent as much as 15% of its weight.

DOW CHEMICAL CO.

WORLD

Accession no.831415

Item 23

European Plastics News

28, No.5, May 2001, p.53

VINYL FRONTIER

Vink D

The PVC industry is stepping up its interest in chemical recycling. Chemical recycling into feedstock raw materials is becoming more popular, especially for mixed waste with high PVC content. Sources include floor coverings, cables, artificial leather and end-of-life vehicles. The European Council of Vinyl Manufacturers, the Vinyl Institute of the US and plant manufacturer Linde have invested around Euro3.3m in a pilot plant that will convert waste with high PVC content into raw materials for the chemicals and plastics industries. The 2,000 t/y plant has been built at Solvay's site in Tavaux, France, and uses a slag gasification process.

EUROPEAN COUNCIL OF VINYL MANUFACTURERS
EUROPEAN COMMUNITY; EUROPEAN UNION; FRANCE; WESTERN EUROPE
Accession no.831401

Item 24
European Plastics News
28, No.5, May 2001, p.37-8
NATURAL SELECTION

The renewability and environmentally friendly qualities of natural fibres have led to their increased use in composite applications, particularly in the European automotive industry. Only a small fraction of composite waste can be recycled, the rest must be incinerated. Incorporating natural fibres in the composites can reduce the impact of incineration, because they only release as much CO2 as the plant absorbed when it was growing. Natural fibres can also be chemically recycled back to raw materials. However, European environmental legislation is threatening the use of natural fibres in composites. The ELV directive sets recycling quotas of 80% by 2007 and 85% by 2015, allowing feedstock recycling to raw materials only in exceptional cases. AKV has asked the German federal environment ministry to interpret the directive to allow feedstock recycling and energy recovery as equivalent to recycling for renewable materials.
WESTERN EUROPE-GENERAL
Accession no.831395

Item 25
Plast' 21
No.102, May 2001, p.77-80
Spanish
PYROLYSIS AS A METHOD FOR THE RECYCLING OF COMPOSITE MATERIALS
de Marco I; Torres A; Laresgoiti M F; Caballero B M; Cabrero M A; Gonzalez A; Cambra J F; Legarreta J A; Chomon M J; Gondra K
Pais Vasco,Universidad

A pyrolysis technique was investigated as a method for the chemical recycling of glass fibre-reinforced unsaturated polyester SMC composites. The process yielded liquid products and gases and also a solid residue formed in the pyrolysis of glass fibres and fillers. The solid residue was used as a reinforcement/filler in unsaturated polyester BMC composites, and the influence on mechanical properties was studied in comparison with BMC prepared entirely from virgin materials.
GAIKER
EUROPEAN COMMUNITY; EUROPEAN UNION; SPAIN; WESTERN EUROPE
Accession no.831364

Item 26
POLYPROPYLENE IN PACKAGING. Proceedings of a one-day seminar, held Shawbury, 25th June 1998. Shawbury, 1998, Paper 4, pp.2. 012
RECOVERY AND RE-USE OF PLASTICS PACKAGING
Smith B R
Valuplast Ltd.
(Rapra Technology Ltd.)

The situation with respect to the recycling and re-use of plastics packaging materials in the UK is examined with reference to data complied by Valuplast Ltd. covering UK plastics recycling achievements to data and forecasts. Data covers the periods 1995 to 2001. Recycling by mechanical methods, polymer cracking, gasification, blast furnace, and feedstock recycling is examined with reference to current and proposed installations. The role of Valuplast is noted. The information is presented in the form of a series of slides.
EUROPEAN COMMUNITY; EUROPEAN UNION; UK; WESTERN EUROPE
Accession no.830719

Item 27
Journal of Applied Polymer Science
81, No.9, 29th August 2001, p.2249-52
MECHANOCHEMICAL SOLID-PHASE REACTION BETWEEN POLYVINYLIDENE FLUORIDE AND SODIUM HYDROXIDE
Qiwu Zhang; Jinfeng Lu; Saito F; Baron M
Tohoku,University; Ecole des Mines d'Albi Carmaux

A planetary ball mill was used to grind a mixture of PVDF and sodium hydroxide. The grinding stimulated the solid phase reaction to form sodium fluoride and organic compounds. About 90% of the fluorine in the PVDF sample could be converted into sodium fluoride by this process. Water soluble and insoluble oxygen-containing compounds were formed in the organic phase via substitution of fluorine in PVDF by the OH base. These organic compounds were potentially useful as fuel, whereas the sodium fluoride was potentially useful for fluorine recycling. 13 refs.
EUROPEAN COMMUNITY; EUROPEAN UNION; FRANCE; JAPAN; WESTERN EUROPE
Accession no.828774

Item 28
Journal of Applied Polymer Science
81, No.9, 29th August 2001, p.2102-8
DEPOLYMERIZATION OF POLYETHYLENETEREPHTHALATE IN SUPERCRITICAL METHANOL
Kim B-K; Hwang G-C; Bae S-Y; Yi S-C; Kumazawa H
Hanyang,University

The depolymerisation of PETP in supercritical methanol was caried out using a batch-type autoclave reactor. The conversion and yield of dimethyl terephthalate (DMT) increased with rising temperature. The yield of DMT exceeded 50% above 280C and the final yield of DMT at 300 and 310C reached 97.0% and 97.7% respectively. the yield of ethylene glycol was slightly lower than that of DMT. the yield of DMT increased markedly when the methanol density was 0.08 g/cc and levelled off at higher densities. A kinetic model to simulate the depolymerisation of PETP in supercritical methanol was suggested. The values of the forward reaction rate constants at different temperatures were determined by comparing the observed time dependence with that calculated by the proposed model. The activation energy was found to be 49.9 kJ/mol, which was close to a previously published value of 55.7 kJ/mol. 4 refs.

KOREA

Accession no.828757

Item 29

Kunststoffe Plast Europe

91, No.8, Aug. 2001, p.44-6

THE PVC LOOP CLOSED

Yernaux J-M; Saffert R

Solvay SA; SolVin

The development of a new recycling process, which selectively dissolves the PVC in composite products, is reported. The process, developed by Solvay/SolVin and called Vinyloop, produces a clean PVC powder, which may be used without further processing and in some cases may be employed in the original application (closed-loop recycling). Scale-up of the process to industrial production level is briefly discussed, the priorities structure for collecting the material is illustrated using floor covering, as an example, and several Vinyloop projects, which have been started up, are indicated. (Kunststoffe, 91, No.8, 2001, p.118-21)

BELGIUM; EUROPEAN COMMUNITY; EUROPEAN UNION; GERMANY; WESTERN EUROPE

Accession no.828202

Item 30

Antec 2001.Conference proceedings.

Dallas, Texas, 6th-10th May, 2001, paper 389

POLYESTER POLYOLS FOR POLYURETHANES FROM RECYCLED PET

Rossi P; Kosior E; Iovenitti P; Massod S; Sbarski I

Swinburne,University of Technology; Visy Plastics Pty.Ltd.

(SPE)

Post-consumer poly(ethylene terephthalate) (PETP) bottle waste flake was depolymerised using triethylene glycol at a temperature of 230 C. The rate of depolymerisation was monitored by gel permeation chromatography and rheology measurements at 50 C using a cone plate viscometer. Digestion was completed in 6 h. Saturated polyester polyols was produced by reacting the glycolised PETP with adipic acid and a small amount of pentarythritol branching agent. 8 refs.

AUSTRALIA

Accession no.826605

Item 31

International Journal of Polymeric Materials

49, No.2, 2001, p.205-15

TWO PACK POLYURETHANE COATINGS FROM PET WASTE AND BIOLOGICAL MATERIALS

Patel J V; Soni P K; Sinha V

Sardar Patel University

The feasibility of using starch-derived glycol-glycosides for depolymerisation of PETP waste recycled from post-consumer soft drink bottles and the use of the depolymerised oligomers for synthesis of liquid polyester polyol by reacting these oligomers with soya fatty acids for PU coatings were studied. The performance properties revealed that the coatings formed could successfully be used for applications in industrial maintenance finishes and were significantly affected by the amount of PETP. 22 refs.

INDIA

Accession no.825133

Item 32

High Performance Polymers

13, No.2, June 2001, p.S365-71

SURFACTANTS BASED ON RECYCLED POLYETHYLENE TEREPHTHALATE FOR BREAKING WATER-IN-OIL EMULSIONS

Abdel-Azim A-A A; El-Sukkary M M A

Egyptian Petroleum Research Institute

Details are given of the depolymerisation of PETP waste with propylene glycol to an oligoester. The glycolised product was reacted with polyethylene glycol of different molecular weights to form surfactants having different hydrophile-lipophile balances. The interfacial tension at the aqueous-benzene interface was determined. 24 refs.

EGYPT

Accession no.823896

Item 33

Plastics Technology

47, No.8, Aug.2001, p.58-61

NEW WAYS TO SALVAGE PLASTIC WASTE

Schut J H

Western Europe recovered 32% of its post-consumer plastic waste in 1999. However, it is claimed that only 11% of total plastic waste is actually recycled. Some 21% is burned for energy. Several new recycling technologies are being commercialised for the first time, holding the promise of signficantly higher levels of plastics reuse. Vinyloop, developed by Solvay, is a solvent-based separation technology which recovers PVC from wire chop after the copper is removed. A second solvent-based recycling technology, developed by Delphi Automotive Systems, removes PVC from whole automotive wire harnesses. Another new European separation approach uses a combination of heat and spinning to separate plastics that are amalgamated with each other, like regrind of overmoulded auto parts. A new near-infrared flake-identification system developed by LLA Laser Labor Adlershof is said to be the first infrared device that can correctly identify the resins in black plastic chips.

WESTERN EUROPE-GENERAL

Accession no.823337

Item 34

Patent Number: EP 1104753 A1 20010606

PROCESS FOR RECOVERING MONOMERIC UNITS OF A NYLON FROM WHOLE CARPET

Courage A J F M; Houben M J A; Mertens M H M; Raets L J G

DSM NV

Disclosed is a process for recovering monomeric units of a nylon from whole carpet composed of fibres of the nylon and a backing composed of non-nylon components, the fibres being bound to the backing and the carpet containing between 15 and 35 wt.% of the nylon. It involves the steps of a) mechanically separating the whole carpet into a carpet mixture, which contains between 35 and 55 wt.% of nylon, and a depleted carpet mixture, and exposing the carpet mixture to conditions under which depolymerisation of the nylon is effected.

EUROPEAN COMMUNITY; EUROPEAN UNION; NETHERLANDS; WESTERN EUROPE; WESTERN EUROPE-GENERAL

Accession no.820288

Item 35

Polymer Degradation and Stability

70, No.1, 2000, p.97-102

CATALYTIC DEGRADATION OF IGH DENSITY POLYETHYLENE AND POLYPROPYLENE INTO LIQUID FUEL IN A POWDER-PARTICLE FLUIDIZED BED

Guohua Luo; Suto T; Yasu S; Kato K

Gunma,University

The catalytic degradation of HDPE and PP was carried out in a powder-particle fluidised bed with an inside diameter of 25 mm. Two catalysts were used as the medium fluidisation particles. These were F9 (with a composition of silica:alumina:sodium oxide of 32:48:20 wt.percent) and silica/alumina (SA, with a composition of silica:alumina of 71:29 wt.percent). Liquid fuel was produced in high yield. SA produced a higher yield of liquid fuel nd a more valuable gas product between temperatures of 400-550C. Over 86 wt.percent conversion of liquid fuel containing hydrocarbons from C-5 to C-11 was produced from HDPE and PP below 500C with a residue yield of less than 8 wt.percent.The gas product, under appropriate reaction conditions, contained over 59 wt.percent ethylene and propylene. 13 refs.

JAPAN

Accession no.819234

Item 36

Polymer Degradation and Stability

72, No.3, 2001, p.469-91

VACUUM PYROLYSIS OF COMMINGLED PLASTICS CONTAINING PVC. PART I KINETIC STUDY

Miranda R; Jin Yang; Roy C; Vasile C

Laval,Universite; Institut Pyrovac Inc.; Petru Poni,Institute of Macromolecular Chemistry

The thermal decomposition behaviour of commingled plastics during incineration was evaluated with particular emphasis on the influence of chlorine released from PVC during pyrolysis. The principal polymers found in municipal plastics waste were evaluated; HDPE, LDPE, PP, PS and PVC. This initial program studied the pyrolysis kinetics of the five individual polymers, mixed polymers without PVC and mixed polymers including PVC. Two experimental approaches were adopted; comparing the decomposition curves for the mixed and individual polymers, and a comparison of the kinetic parameters for each material. Samples were heated at varying heating rates both under a vacuum and in a nitrogen atmosphere. Thermogravimetric weight loss and weight loss derivative curves were recorded against time. The results indicated that some interactions occurred between the plastics materials during pyrolysis mainly above 375 deg. C as their individual decomposition rates were significantly altered. However the chlorine from the PVC was released almost completely below 375 deg. C. Detailed experimental procedures and results are given. 73 refs.

CANADA; EASTERN EUROPE; RUMANIA

Accession no.818936

Item 37

Materie Plastiche ed Elastomeri

65, No.10, Oct.2000, p.728-30

Italian

MAKE THE MOST OF IT, BURN IT

Calato F

Incineration with energy recovery is examined as a means for the disposal of plastics waste, and data are presented for the calorific values of a number of materials. Chemical recycling techniques are also briefly reviewed.

EUROPEAN COMMUNITY; EUROPEAN UNION; ITALY; WESTERN EUROPE; WESTERN EUROPE-GENERAL

Accession no.818428

Item 38
Revista de Plasticos Modernos
80, No.531, Sept.2000, p.276-8
Spanish
ENERGY RECOVERY FROM SCRAP POLYETHYLENE GREENHOUSE FILMS
Rodriguez J
Repsol Quimica

Results are presented of a study undertaken at a power station in Almeria, Spain, to assess the effectiveness of a co-combustion process in the disposal with energy recovery of scrap PE greenhouse covering films.

REPSOL-YPF; ENDESA; APME; DOW CHEMICAL IBERICA SA; CIBA SPECIALTY CHEMICALS; RECICLADOS NIJAR SL
BELGIUM; EUROPEAN COMMUNITY; EUROPEAN UNION; SPAIN; WESTERN EUROPE
Accession no.818393

Item 39
Asian Plastics News
June 2001, p.31
IHI MAKES FUEL FROM USED PLASTICS

Ishikawajima-Harima Heavy Industries has started making solid fuel from used plastics at a plant in Chita, Aichi prefecture in central Japan. The company has for some time now been making recycled plastics fuel from used plastics collected from home appliance makers and print shops. IHI recently boosted its processing capacity to 1,000 t/m and intends to add more when this operation gets into full swing. The company is also set to begin intermediate processing of used plastics containers and consumer electronic products at a former factory site in Nagoya, near the Chita facility.

ISHIKAWAJIMA-HARIMA HEAVY INDUSTRIES CO.LTD.
JAPAN
Accession no.816684

Item 40
Shawbury, Rapra Technology Ltd., 2000, pp.130. 30 cms., 1/6/01. Rapra Industry Analysis Series
PLASTICS PRODUCT RECYCLING
Hulse S
Rapra Technology Ltd.

Plastics products recycling in Europe is examined, with reference to various end-use industry, factors affecting the recycling efforts, consumption of plastics by the sector, and examples of recycling schemes by country and industry. Legislation relevant to each country is examined, with details of recycling organisations and compliance schemes. The recovery of plastics waste by mechanical and chemical methods, and for energy recovery are described, including techniques for the separation of mixed plastics wastes. End-use sectors examined include packaging, automotive, electrical and electronic, building and construction, and agriculture and horticulture. Environmental and economic implications arising from plastics waste are also considered.

WESTERN EUROPE
Accession no.815478

Item 41
Patent Number: US 6160031 A1 20001212
PROCESS FOR DECOMPOSING A POLYMER TO ITS MONOMER OR MONOMERS
Poree I D; Cameron K P; Bloem J A; Schlosser F D; McGowan A
AECI Ltd.

This involves subjecting a polymer, such as PMMA, in solid, gel, partially molten or molten form to microwave heating for a time and at a temperature sufficient to decompose the polymer to produce a monomer or monomers in gaseous, liquid or solid form, without substantial decomposition of the monomer or monomers, and recovering at least one of the monomer or monomers. The monomer or monomers may then be reused for polymerisation.

SOUTH AFRICA; USA
Accession no.812779

Item 42
Kunststoffe Plast Europe
91, No.2, Feb.2001, p.29-31
English; German
CONTRADICTION IN TERMS?
Wiedemann P
Wipag Polymertechnik

With the adoption of the EU End of Life Vehicle Directive, the subject of recycling has again moved into the foreground. In addition to the possibility of cost-free take-back for the last owner, the quotas to be achieved are of vital importance. From 2006: more than 80% material recycling (including reuse) and total recycling more than 85% total recycling (reuse, material recycling, thermal energy recovery). From 2015: more than 85% material recycling (including reuse) and greater than 95% total recycling (reuse, material recycling, thermal energy recovery). A point of considerable importance that is often underestimated is the stipulation that type approval for a vehicle on the market after 1st January 2005 will in future only be granted if it can be demonstrated that the recycling quotas of 2015 can be fulfilled. (Translated from Kunstoffe 91, 2001, 2, p.67-9).

EU; EUROPEAN COMMUNITY; EUROPEAN UNION; GERMANY; WESTERN EUROPE; WESTERN EUROPE-GENERAL
Accession no.809453

Item 43
Patent Number: US 6140428 A1 20001031
SIMULTANEOUS PRODUCTION OF DICARBOXYLIC ACIDS AND DIAMINES BY SPLITTING POLYAMIDES INTO THEIR MONOMERIC CONSTITUENTS
Seeliger U; Mueller W F; Heimann F; Huber G; Habermann W; Voss H; Siegel H
BASF AG

Disclosed is a process for the simultaneous production of dicarboxylic acids and diamines from a) polymers based on polyamides of dicarboxylic acids or their derivatives with diamines or b) compositions containing essentially such polymers. It involves treating these polymers or compounds with a base in alcoholic medium and subsequently converting the resulting dicarboxylate salts electrochemically into the corresponding dicarboxylic acids and bases.

EUROPEAN COMMUNITY; EUROPEAN UNION; GERMANY; USA; WESTERN EUROPE

Accession no.809221

Item 44
Patent Number: US 6136869 A1 20001024
DEPOLYMERIZATION PROCESS FOR RECYCLING POLYESTERS
Ekart M P; Pell T M; Cornell D D; Shackelford D B
Eastman Chemical Co.

A process for recovering suitable polyester feedstock material from recycled polyester includes the steps of depolymerising the polyester into its component ester monomers and half-esters; separating the monomers and half-esters from other secondary materials; and mixing the component ester monomers and half-esters with additional monomers to produce a low molecular weight polyester. In a preferred embodiment of the invention the recycled polyester is contacted with a component monomer or oligomer thereof to liquefy the polyester before the polyester is depolymerised. The invention further provides an apparatus for carrying out the process. The apparatus includes a first reactor for depolymerising the recycled polyester, a separator for separating ester monomeric components and half-esters from secondary materials produced in the first reactor, and a second reactor for producing a low molecular weight polyester from the liquefied separator products.

USA

Accession no.808944

Item 45
Industrial & Engineering Chemistry Research
40, No.2, 24th Jan.2001, p.509-14
KINETICS OF DEPOLYMERIZATION OF POLY(ETHYLENE TEREPHTHALATE) IN A POTASSIUM HYDROXIDE SOLUTION
Ben-Zu Wan; Chih-Yu Kao; Wu-Hsun Cheng
Taiwan,National University; Taiwan,Chang Gung University of Technology

The hydrolytic depolymerisation of PETP in stirred potassium hydroxide solution was investigated. It was found that the depolymerisation reaction rate in a KOH solution was much more rapid than that in a neutral water solution. The correlation between the yield of product and the conversion of PETP showed that the main alkaline hydrolysis of PETP linkages was through a mechanism of chain-end scission. The result of kinetic analysis showed that the reaction rate was first order with respect to the concentration of KOH and to the concentration of PETP solids, respectively. This indicated that the ester linkages in PETP were hydrolysed sequentially. The activation energy for the depolymerisation of solid PETP in a KOH solution was 69 kJ/mol and the Arrhenius constant was 419 L/min/sq cm. 21 refs.

TAIWAN

Accession no.807152

Item 46
Plastics and Rubber Weekly
23rd Feb.2001, p.12
DANISH ROAD TO RECOVERY
Hague C

More than 80% of British waste is dumped in landfill sites. In Denmark, the figure is 4%. Under proposed revisions to the 1994 packaging and packaging waste directive, by 2006 the UK will have to recycle 20% of plastics packaging waste. Currently the UK recycles 8%, with recycling capacity stagnant. The BPF organised a study trip to Denmark to show how an integrated waste management scheme can be implemented. Copenhagen does not in fact recycle any plastics. All polymers stripped from the waste stream are burned for energy recovery in incinerators. There are 32 incinerators in Denmark serving a population of 5 million. In the UK, there are 12 for 58 million people.

DENMARK; EUROPEAN COMMUNITY; EUROPEAN UNION; SCANDINAVIA; UK; WESTERN EUROPE

Accession no.806826

Item 47
Antec 2000.Conference proceedings.
Orlando, Fl., 7th-11th May, 2000, paper 581
MUNICIPAL PLASTIC WASTE: ALTERNATIVES FOR RECYCLING WITH PROFIT
Eulalio A C; Capiati N J; Barbosa S E
Bahia Blanca,Universidad Nacional del Sur
(SPE)

The recycling of plastic waste, and the final disposal is considered from an energy point of view, and compared with the energy requirements for producing the virgin material. When the energy gains from incineration are deducted from the energy requirement to produce the replacement polymer, and compared with the energy requirements for recycling, a 70% energy saving may be achieved by recycling polyethylene and polypropylene, and 50% when recycling poly(ethylene terephthalate). The energy values for commingled polymers may be calculated by the rule of mixtures. 8 refs.

ARGENTINA

Accession no.803878

Item 48

Polymer Recycling
5, No.1, 1999/2000, p.15-22

UNSATURATED POLYESTER RESINS BASED ON THE PET WASTE GLYCOLYSIS PRODUCTS BY ETHYLENE, PROPYLENE AND DIETHYLENE GLYCOLS AND THEIR MIXTURES

Viksne A; Kalnins M; Rence L; Berzina R
Riga,Technical University

Recycling of PETP soft drink bottles was carried out by depolymerisation through glycolysis with ethylene and propylene glycol mixtures with diethylene glycol. The glycolysed products were reacted with maleic anhydride and mixed with styrene or tri(ethylene glycol) dimethacrylate monomers to produce unsaturated polyesters suitable for producing varnishes and paints. The curing behaviour of these resins in the presence of different initiators was studied by mans of DSC, and the influence of curing conditions on the degree of crosslinking and residual enthalpy was analysed. Processing characteristics such as viscosity, exotherm temperatures of curing, compatibility of unsaturated polyester resins with monomers were also studied with respect to the amount and type of reactive monomers and initiators. Properties of the varnish and paint coatings obtained were determined, with particular respect to hardness, impact resistance and elasticity. 8 refs.

LATVIA

Accession no.789679

Item 49

Shawbury, Rapra Technology Ltd., 2000, pp.144. 29 cms. Rapra Review Report No.121, Vol. 11, Number 1, 2000. NALOAN

POLYAMIDES AS ENGINEERING THERMOPLASTIC MATERIALS

Page I B
BIP Ltd.
Edited by: Dolbey R
(Rapra Technology Ltd.)

This review is concerned with the engineering thermoplastic uses of polyamide materials in injection moulding and extrusion applications. Types of polyamides are described, and their key properties are considered. Commercial applications in the automotive, electrical/electronic, engineering and construction, and packaging industries are discussed. Polyamide processing is examined with reference to drying, injection moulding, extrusion and post processing operations. Modification of properties by compounding is described with reference to the use of flame retardants, plasticisers, stabilisers, lubricants, nucleants, fillers and reinforcements, and also blending to produce toughened and rigid engineering blends. Recycling by mechanical and chemical techniques is covered, including fibre reprocessing. 516 refs.

EUROPEAN COMMUNITY; EUROPEAN UNION; UK; WESTERN EUROPE

Accession no.787651

Item 50

Journal of Applied Polymer Science
77, No.14, 29th Sept.2000, p.3228-33

DEPOLYMERIZATION OF POLY(BUTYLENE TEREPHTHALATE) USING HIGH-TEMPERATURE AND HIGH-PRESSURE METHANOL

Shibata M; Masuda T; Yosomiya R; Meng Ling-Hui
Chiba,Institute of Technology; Harbin,Institute of Technology

PBTP was depolymerised in excess methanol under high temperature (473-523 K) and high pressure (4-14 MPa) conditions. Depolymerisation was carried out at 483 K and 4-12 MPa, and at 513 K and 6-14 MPa. The temperature had a great effect on the depolymerisation rate, but the reaction pressure did not. Under the former conditions, depolymerisation took over 80 min, but only about 20 min under the latter conditions. The se results showed that the supercritical state of methanol was not a key factor in the depolymerisation reaction. A kinetic study of the reaction at 473-523 K and 12 MPa showed that the decomposition rate constant of PBTP increased dramatically when the reaction temperature was higher than the melting point of PBTP (500 K). This indicated that partial miscibility of the molten PBTP and methanol was an important factor for the short-time depolymerisation. 9 refs.

CHINA; JAPAN

Accession no.784903

Item 51

Journal of Applied Polymer Science
77, No.12, 19th Sept. 2000, p.2646-56

DESAMINATED GLYCOLYSIS OF WATER-BLOWN RIGID POLYURETHANE FOAMS

Lee J Y; Kim D
Sung Kyun Kwan University

Glycolysis of rigid, MDI-based PU foams was carried out using three different glycols (ethylene glycol, propylene glycol and diethylene glycol) and the influence of glycol solvent on the kinetics of the reaction investigated. Glycolysates were reacted with butyl glycidyl ether to convert toxic aromatic amines to polyols, which were identified by means of gel permeation chromatography. The

PU foams were reprepared from virgin and recycled polyol mixtures and their morphological and physical properties, such as density, thermal conductivity and flexural strength, compared. 15 refs.

SOUTH KOREA

Accession no.784345

Item 52

Polymer

41, No.18, 2000, p.6749-53

CHEMICAL CONVERSION OF POLY(CARBONATE) TO BIS(HYDROXYETHYL) ETHER OF BISPHENOL A. AN APPROACH TO THE CHEMICAL RECYCLING OF PLASTIC WASTES AS MONOMERS

Oku A; Tanaka S; Hata S
Kyoto,Institute of Technology

A method of converting polycarbonate (PC) to bishydroxyethyl ether of bisphenol A (BHE-BPA) was studied, with a view to recycling PC plastic wastes. Treating PC in ethylene glycol with a catalytic amount of sodium hydroxide produced the monohydroxyethyl ether of bisphenol A (MHE-BPA, 42%), BHE-BPA (11%) and BPA (42%). BHE-BPA was produced quantitatively when 1.6 mol. equiv. ethylene carbonate was added to this reaction system. The reaction of BPA with EC produced both BHE-BPA and MHE-BPA, indicating that ethylene carbonate was formed as an intermediate in the base catalysed reaction of PC with ethylene glycol. A large proportion of this ethylene carbonate formed from PC was, however, lost by decarboxylation so additional ethylene carbonate must be provided for the quantitative preparation of BHE-BPA. 12 refs.

JAPAN

Accession no.782228

Item 53

Iranian Polymer Journal

9,No.1,Jan.2000,p.37-40

CHEMICAL RECYCLING OF POLYETHYLENE TEREPHTHALATE

Mehrabzadeh M; Shodjaei S T; Khosravi M
Iran,Polymer Institute; Tehran,Islamic Azad University

Polyethylene terephthalate (PET) is one of the most important commercial thermoplastic polyesters, which has been on the market since 1977 and is widely used in both industrial and household applications. Under specific conditions, plastics can be converted into their primary components for use in other chemical processes by chemical recycling. PET is a thermoplastic, and so recycling by chemical methods, which converts it into primary components, can be achieved. This study examines the optimal routes of the existing chemical methods. For chemical recycling, acidic hydrolysis is used and PET is converted into terephthalic acid (TPA) and ethylene glycol (EG). Effective factors in the decomposition and the yield such as acid concentration, time, temperature and PET particle size were investigated. Characterisation of the products was carried out by FTIR, and the effect of reaction time on PET particles evaluated by scanning electron microscopy. 6 refs.

IRAN

Accession no.780247

Item 54

Macromolecular Symposia

Vol.152, March 2000, p.191-9

RECYCLING OF MIXED PLASTICS BY PYROLYSIS IN A FLUIDISED BED

Kaminsky W; Schmidt H; Simon C M
Hamburg,University

The thermal cracking of a light fraction of mixed plastics waste was carried out in a fluidised bed reactor and the fractions obtained were analysed by elemental analysis, gas chromatography and ashing. The main components of the waste were PE and PP with a small amount of PS and the bed was fluidised by pyrolysis gas, nitrogen or preheated steam. Experiments conducted at different temperatures and residence times were compared by calculating the crack severity for each experiment. The results obtained revealed that the amounts of ethene and propene obtained by pyrolysis with steam were comparable with those obtained using a commercial steam cracker.

EUROPEAN COMMUNITY; EUROPEAN UNION; GERMANY; WESTERN EUROPE

Accession no.778696

Item 55

Progress in Rubber and Plastics Technology

16, No.1, March 2000, p.61-8

CHEMICAL RECYCLING OF WASTE SATURATED POLYESTERS AND URETHANE POLYMERS TO YIELD RAW MATERIALS FOR THE PRODUCTION OF POLYURETHANES

Kacperski M; Spychaj T
Szczecin,Polytechnic

Reduction of the amount of waste sent to refuse dumps is now a very important problem. The rising costs of storing waste products, and public protests in connection with the construction of new dumps, are prompting scientists and Polish workers in industry to undertake research into reduction of the amount of waste sent to refuse dumps. It appears that the Wastes Act, which came into force on 1 January 1998, will be an impetus for increasing the intensity of this work. In Poland the proportion of plastics in the total mass of municipal waste has increased considerably during the last decade. Used waste plastics can be re-used, once they have been separated from the rest of the waste and cleaned. There are two categories of methods of management of waste plastics: processing the waste products without altering their chemical structure, for example in injection moulding or extrusion moulding processes (material recycling); and processing of waste products with simultaneous alteration of their chemical structure (chemical recycling). The possibility of utilising waste saturated polyesters and PUs as raw materials in the production of substrates for the manufacture of PU plastics are discussed. 29 refs.

EASTERN EUROPE; POLAND

Accession no.778132

Item 56
Macplas International
No.5, May 2000, p.59-61
ECO-EFFICIENCY OF PACKAGING WASTE RECOVERY
Mayne N
APME

A study of the eco-efficiency of the recovery of packaging waste was commissioned by APME and undertaken by the Dutch scientific research institute, the TNO. This article provides information on this study, under sections entitled: model for the study, waste scenarios, environmental and economic impact, eco-efficiency of waste management, and finally, main conclusions.

TNO; DSD
EUROPE-GENERAL; EUROPEAN COMMUNITY; EUROPEAN UNION; GERMANY; NETHERLANDS; WESTERN EUROPE

Accession no.777450

Item 57
ENDS Report
No.303, April 2000, p.29-31
WASTE PROBLEMS LOOM LARGER FOR PVC

This article presents details of the gloomy picture of the environmental impacts associated with PVC waste which has been painted by four PVC waste management studies carried out for the European Commission. The studies cover mechanical recycling, feedstock recycling, behaviour in landfill, and the influence of PVC on incinerator flue gas cleaning residues.

EUROPEAN COMMISSION
EU; EUROPEAN COMMUNITY; EUROPEAN UNION; UK; WESTERN EUROPE; WESTERN EUROPE-GENERAL

Accession no.777317

Item 58
Chemical Engineering
107, No.6, June 2000, p.41
NEW PLASTICS-RECYCLING MANDATES
Crabb C

The current and future levels of plastic packaging recycling is discussed with reference to the imminent updating of the 1994 Packaging and Packaging Waste Directive. Data are included relating to current recovery levels by country for western Europe, for mechanical and feedstock recycling activities, and the need to balance recycling and energy recovery efforts is considered.

EUROPEAN COMMISSION
EU; EUROPEAN COMMUNITY; EUROPEAN UNION; WESTERN EUROPE-GENERAL

Accession no.776971

Item 59
Plastics Newsletter
No.2, 2000, p.29-30
FIRST LARGE-SCALE CARPET RECYCLING PLANT

Evergreen Nylon Recycling, a DSM/AlliedSignal joint venture, opened in November 1999. The plant will convert over 90,000 t/y of nylon 6 carpet waste into the raw material for nylon, caprolactam. At the moment 450,000 tonnes of nylon 6 carpet waste is transported to US landfills each year. A fifth of this will be recycled in the new plant. The 45,000 tonnes of caprolactam that Evergreen Nylon Recycling will produce every year can be used to make nylon 6 of the same quality and with the same properties as virgin nylon 6.

EVERGREEN NYLON RECYCLING LLC
USA

Accession no.775771

Item 60
Polymer Engineering and Science
40, No.4, April 2000, p.979-84
MODEL STUDY FOR THE RECOVERY OF POLYAMIDES USING THE DISSOLUTION/ REPRECIPITATION TECHNIQUE
Papaspyrides C D; Kartalis C N
Athens,National Technical University

Dissolution/reprecipitation processes were evaluated for the recycling of poly-epsilon-caprolactam (PA6) and polyhexamethyleneadipamide (PA66). The process involved solution of the polyamide in an appropriate solvent, precipitation by the addition of a non-solvent, and recovery of the polymer by washing and drying. Dimethylsulphoxide was used as the solvent for PA6, and formic acid for PA66, and methylethylketone was used as the non-solvent for both polymers. The recycled polymers were evaluated by determination of molecular weight, crystallinity and grain size. Excellent recoveries were achieved, with no deterioration in the polymer properties. 33 refs.

EUROPEAN COMMUNITY; EUROPEAN UNION; GREECE; WESTERN EUROPE

Accession no.775666

Item 61
Polymer
41, No.11, 2000, p.4361-5
RAPID MICROWAVE INDUCED DEPOLYMERISATION OF POLYAMIDE 6
Klun U; Krzan A

Ljubljana,National Institute of Chemistry

Polyamide 6 depolymerisation was performed using microwaves as the energy source for the acid catalysed hydrolysis, with phosphoric acid as the catalyst. The product mixture was analysed by chromatographic and spectroscopic methods. 19 refs.

SLOVENIA

Accession no.774158

Item 62

Journal of Thermoplastic Composite Materials
13, No.2, March 2000, p.92-101

TERTIARY RECYCLING OF AUTOMOTIVE PLASTICS AND COMPOSITES

Allred R E; Busselle L D
Adherent Technologies Inc.

This paper summarises an initial feasibility on recyling scrap automotive plastics and composites using a catalytic conversion process. The characterisation of hydrocarbon products is presented for sheet moulding compound (SMC), auto shredder residue (ASR) and reinforced polypropylene (R-PP) materials and mixtures of body panels. Gas chromatography and scanning electron microscopy is used for the material characterisation. 26 refs.

USA

Accession no.773081

Item 63

Composites Science & Technology
60, No.4, 2000, p.509-23

FLUIDISED-BED PROCESS FOR THE RECOVERY OF GLASS FIBRES FROM SCRAP THERMOSET COMPOSITES

Pickering S J; Kelly R M; Kennerley J R; Rudd C D; Fenwick N J
Nottingham,University

A fluidised bed combustion process was developed for treatment of thermoset composites in the form of process scrap or end-of-life components. The process was shown to be capable of dealing with contaminated scrap of variable composition and to provide useful outputs in the form of recovered fibres, particulate materials and heat. Comminuted feeds were decomposed at a bed temp. of 450C and a fluidising velocity of 1.3 m/s. Fibres with mean lengths of up to 5 mm were collected at purities of up to 80% by using a rotating sieve separator. Shorter fibres were collected with the particulate mineral fillers. The TS of recovered E-glass fibres was reduced by up to 50% although this depended on the thermal history within the process. Fibre modulus was relatively unaffected by the exposure to high temps. Recovered fibres were successfully reused in dough moulding compound formulations and veil products. A simple economic model indicated that the process would break even at a throughput of about 9000 tonnes scrap composite per year. 23 refs.

EUROPEAN COMMUNITY; EUROPEAN UNION; UK; WESTERN EUROPE

Accession no.772526

Item 64

Canadian Plastics
58, No.4, April 2000, p.8

NYLON SUPPLIERS TAKE LEAD IN CARPET RECYCLING

A number of nylon manufacturers are or are about to recycle carpets in Canada, the USA and Europe. DuPont is planning to build a demonstration plant in Maitland, Ont. for the recycling of nylon 66 and nylon 6 carpet and Evergreen Nylon Recycling operates a commercial-scale, closed-loop recycling plant in Augusta, GA. BASF converts post-consumer carpet made from its own Zeftron nylon 6 into caprolactam at a plant in Arnipor, Ont. and Rhodia depolymerises nylon 6 waste at various European plants.

BASF CANADA INC.; DUPONT CANADA INC.; EVERGREEN NYLON RECYCLING LLC; RHODIA ENGINEERING PLASTICS

CANADA

Accession no.772446

Item 65

Industrial & Engineering Chemistry Research
39, No.5, May 2000, p.1198-202

CATALYTIC DEGRADATION OF HIGH-DENSITY POLYETHYLENE OVER DIFFERENT ZEOLITIC STRUCTURES

Manos G; Garforth A; Dwyer J
London,University College; Manchester,University

A range of zeolites were studied in their capacity to degrade high-density polyethylene to hydrocarbons. Zeolite structure, especially pore size, was found to directly influence products formed. Medium pore size yielded significantly more olefins as secondary bimolecular reactions are sterically hindered resulting in higher amounts of alkenes as primary products, whereas those with larger pore size gave products of greater saturation. Medium pore sized zeolites also yielded lighter products than those with larger pores. Depending on the zeolite used, products with high fuel values confirm that plastic recycling via catalytic degradation is a promising method. 17refs.

EUROPEAN COMMUNITY; EUROPEAN UNION; UK; WESTERN EUROPE

Accession no.772173

Item 66

Vinyltec '99. Conference proceedings.
Ontario, 12th-14th Oct.1999, p.42-7

ROLE OF PVC IN THE RESOURCE RECOVERY OF HYDROCARBONS FROM MIXED PLASTIC WASTES BY PYROLYSIS

Day M; Shabnavard L; Touchette-Barrette C; Sheehan S E; Cooney J D
Canada,National Research Council
(SPE,Vinyl Div.)

The pyrolysis products obtained from a variety of mixed plastics containing PVC are investigated. While hydrochloric acid is the major chlorinated product produced by PVC pyrolysis, other chlorinated hydrocarbons are produced. However, the composition and yield of these compounds are very much dependent upon the other polymers present in the plastic mixture. In the case of a polymeric waste stream containing inorganic fillers, such as calcium carbonate, the HCl produced by the PVC can be neutralised in situ, leading to the production of inorganic chlorides, alleviating many of the concerns associated with HCl formation. 9 refs.

CANADA

Accession no.769495

Item 67

ARC '99. Conference proceedings.
Detroit, Mi., 9th-11th Nov.1999, p.307-14

PLASTICS AND PROCESS ENGINEERED FUEL (PEF): AN OVERVIEW

Fisher M M; Tomczyk L
Beck R.W.,Inc.; American Plastics Council
(SPE,Plastics Recycling Div.)

Scrap plastics and paper can be made into process engineered fuel (PEF) for co-firing in existing solid fuel boilers. Approximately 20 PEF facilities are currently operating in the USA. PEF systems offer the opportunity to divert significant amounts of materials from disposal that are generally not recycled through traditional means while also conserving fossil fuel resources. PEF is a manufactured product produced in either shredded (fluff) or densified form and has predictable combustion characteristics. PEF prices, fuel characteristics and resulting air emissions can be attractive in comparison to other more conventional fuels. Plastics enhance the energy content of these fuels. The economics of PEF systems are very case specific, and can be favourable under certain conditions. 8 refs.

USA

Accession no.768700

Item 68

ARC '99. Conference proceedings.
Detroit, Mi., 9th-11th Nov.1999, p.275-81

CATALYTIC PROCESS FOR THE RECLAMATION OF CARBON FIBRES FROM CARBON/EPOXY COMPOSITES

Allred R E; Busselle L D; Shoemaker J M
Adherent Technologies Inc.
(SPE,Plastics Recycling Div.)

The polymerisation of thermoset carbon fibre-reinforced epoxy matrix composites is studied to determine the significant reaction parameters, mechanistic reaction model and fibre quality produced by a catalytic reclamation process. This process is designed to recover valuable carbon fibre and an organic fraction from the polymerisation of carbon/epoxy composites. Design of experiments is used to determine a regression model including terms for temperature, time, and agitation to estimate the purity of the carbon fibre produced from the reaction. Depolymerisation of the composites appears to follow a progressive conversion model similar to a solid catalyst reaction during the majority of the reaction. Significant feedstock parameters that will affect the rate of reaction are the surface area available for reaction and the thickness of the composite. The carbon fibres reclaimed from the reaction reach 99.8% carbon values, i.e. 0.2% residual resin, sufficient to meet the market specifications for reuse in conductive moulding compounds. The fibre tensile strength shows 8.6% reduction in strength after reclamation indicating that the process has little damaging effect on the fibre. Potential applications for the recycled fibres include thermoplastic and thermoset moulding compounds and non-woven sheet reinforcements. Economic analysis of a recycling business based on the catalytic depolymerisation process shows that it should be profitable provided that adequate scrap composite feedstock can be obtained. 18 refs.

USA

Accession no.768697

Item 69

ARC '99. Conference proceedings.
Detroit, Mi., 9th-11th Nov.1999, p.265-7

NEW PROCESS TO RECOVER PURE PVC COMPOUNDS FROM COMPOSITES LIKE COATED TEXTILES, FLOORING OR CABLE WASTES

Yernaux J-M
Solvay SA
(SPE,Plastics Recycling Div.)

The Vinyloop process is based on the selective dissolution of PVC used in composites applications like cable insulation, flooring, tarpaulins, blisters, etc. After removal of insoluble parts like metals, rubber or other polymers, the PVC is reprecipitated with all additives by introduction of a non-solvent component which will form with the selective solvent an azeotropic mixture. By using typical conditions, the process is able to recover a pure PVC compound powder ready for use without any additional treatment like melt filtration or a new pelletisation (specific characteristics of the powder are average diameter of 400 microns and bulk density above 600 kg/cub.m). All the solvents used are completely recycled and reused. PVC compounds recovered in the Vinyloop process can be reused in a closed-loop recycling scheme or processed in a large variety of high value applications in calendaring, extrusion or injection.

BELGIUM; EUROPEAN COMMUNITY; EUROPEAN UNION; WESTERN EUROPE

Accession no.768695

Item 70
ARC '99. Conference proceedings.
Detroit, Mi., 9th-11th Nov.1999, p.187-93

THERMOPLASTIC DEPOLYMERISATION: MAKING USEFUL LITTLE ONES OUT OF USED BIG ONES

Cornell D
Eastman Chemical Co.
(SPE,Plastics Recycling Div.)

Commercial plastics polymerisation is akin to making pig's ears out of silk purses, albeit usually useful porcine ears from very worn out handbags. What were once valuable polymers are turned into generally less valuable monomers. The regenerated monomers and small chemicals from polymerisation of post-consumer plastics have no particular moral authority or intrinsic grace compared to chemicals derived from non-recycling sources. To be successful, commercial polymerisation must make economic sense in ways that are understood by those who invest dear money into capital assets.

USA

Accession no.768688

Item 71
ENDS Report
No.302, March 2000, p.44

PARLIAMENT TARGETS PVC IN VOTE ON INCINERATION DIRECTIVE

Amendments to the draft EC Directive on the incineration of PVC are outlined. The European Parliament voted to tighten emission levels for new cement kilns co-incinerating waste, and reduced by one year the proposed exemptions for existing cement kilns burning less than three tonnes of waste per hour. These would have to be upgraded by January 2007 rather than January 2008 as previously agreed. Also agreed was an amendment which would require incinerator operators to take steps to remove large PVC items from the waste stream. Other amendments relating to the combustion of fibrous waste are indicated.

EUROPEAN COMMISSION

EU; EUROPEAN COMMUNITY; EUROPEAN UNION; WESTERN EUROPE-GENERAL

Accession no.768077

Item 72
ENDS Report
No.302, March 2000, p.18

CARPET RECYCLING OFFERS MARKETING ADVANTAGE TO NYLON

The world's first automatic scrap carpet sorting plant is due to open in Germany in late Spring 2000, followed next year by Europe's first carpet recycling plant, also in Germany. The recycling plant will mechanically recycle nylon-6 fibres and depolymerise nylon-6,6. The development of carpet recycling in the EC is said to be largely driven by German legislation requiring manufacturers to examine the possibility of recycling their products and banning carpets from landfill from 2005. Details are given of the new DM 1 million sorting plant which has been built by Carpet Recycling Europe, a company funded by the 87 carpet mills which belong to the European carpet manufacturers association, GuT.

CARPET RECYCLING EUROPE

EUROPEAN COMMUNITY; EUROPEAN UNION; GERMANY; WESTERN EUROPE

Accession no.768075

Item 73
Industrial & Engineering Chemistry Research
39, No.2, Feb.2000, p.245-9

CHEMICAL RECYCLING OF PHENOL RESIN BY SUPERCRITICAL METHANOL

Ozaki J; Djaja S K I; Oya A
Gunma,University

The reaction of phenolic resin in supercritical methanol was studied with variations in temp. from 300 to 420C and in reaction times from 30 to 150 min. The conversion increased rapidly above 350C to give 94% at maximum at 420C and the gas fraction also increased at the same time. When a longer reaction time was used from the reaction at 400C, the conversion increased without giving additional gas product. From the point of view of liquefaction, the longer reaction at lower temp. would give better results. A gas chromatographic study revealed that the liquid product included phenol and its methylated derivatives. The carbon content of the solid product was higher than the initial phenol resin, showing that some carbonisation took place during the reaction. 21 refs.

JAPAN

Accession no.766857

Item 74
ENDS Report
No.301, Feb.2000, p.14

PLASTICS INDUSTRY MISUSES LCA TO LOBBY AGAINST RECYCLING

This article presents the findings of a study carried out for the Association of Plastics Manufacturers in Europe (APME) which favours 85 percent incineration and only 15 percent recycling (over five other waste management scenarios). The article also explains that APME has contravened international standards on life-cycle-assessment (LCA) by promoting the findings of this study before it had been peer-reviewed, and by refusing to publish the full research.

APME; TNO; INTERNATIONAL STANDARDS ORGANISATION; FRIENDS OF THE EARTH

EU; EUROPEAN COMMUNITY; EUROPEAN UNION; NETHERLANDS; UK; WESTERN EUROPE; WESTERN EUROPE-GENERAL

Accession no.765304

Item 75

Kunststoffe Plast Europe

90, No.2, Feb.2000, p.26-8

RECYCLED CFRP AS REINFORCEMENT FOR THERMOPLASTICS

Schubert T; Ehrenstein G

Thermal decomposition of the matrix material offers a simple way of recovering the relatively expensive reinforcing fibres from a fibre-reinforced laminate. The epoxy resin matrix was made to decompose by thermal treatment in air or nitrogen, this treatment allowing the carbon fibres to be recovered without damage.

EUROPEAN COMMUNITY; EUROPEAN UNION; GERMANY; WESTERN EUROPE

Accession no.764597

Item 76

Chemical and Engineering News

78, No.4, 24th Jan. 2000, p.23-4

DUPONT, EVERGREEN TO RECYCLE CARPET FOREVER

Tullo A H

Efforts by several companies in the recycling of nylon carpeting are reported. These companies, notably DuPont, BASF, Evergreen Nylon Recycling, Polyamid 2000, Honeywell and Rhodia, have demonstrated that closed-loop recycling is working. Profitability of the chemical recycling process is discussed, and details are given of individual companies' particular processes, and activities in the field.

DUPONT; EVERGREEN NYLON RECYCLING LLC

USA

Accession no.762056

Item 77

International Polymer Science and Technology

26, No.6, 1999, p.T/93-T/96. (Translation of Polimery, Tworzywa Wielkoczasteczkowe, No.1, 1999, p.1)

CHEMICAL RECYCLING OF WASTE SATURATED POLYESTER AND URETHANE POLYMERS TO YIELD RAW MATERIALS FOR THE PRODUCTION OF POLYURETHANES

Kacperski M; Spychaj T

Techniques are reviewed for the chemical recycling of waste saturated polyesters (PETP and PBTP), and of waste polyurethanes for use in the production of substrates for the manufacture of polyurethane plastics. 29 refs.

EASTERN EUROPE; POLAND

Accession no.762048

Item 78

Polymer Degradation and Stability

67, No.2, Feb.2000, p.285-90

CHEMICAL RECYCLING OF RIGID PVC BY OXYGEN OXIDATION IN NAOH SOLUTIONS AT ELEVATED TEMPERATURES

Yoshioka T; Furukawa K; Okuwaki A

Tohoku,University

Rigid PVC pellets were subjected to oxidative degradation with oxygen in 1 to 25 mol/kg-H2O sodium hydroxide solutions at 150 to 260C in order to establish the fundamental conditions necessary for recycling waste PVC using a novel method. The effects of reaction conditions on weight loss, weight loss kinetics and product yield and distribution were investigated. Major decomposition products were identified as oxalic acid, a mixture of benzenecarboxylic acids and carbon dioxide. The possibility of converting PVC into raw materials, such as carboxylic acids, by chemical recycling is considered. 32 refs.

JAPAN

Accession no.760988

Item 79

Reuse/Recycle

30, No.1, Jan. 2000, p.2-3

WORLD'S LARGEST NYLON RECYCLING PLANT OPENS

The world's largest nylon recycling plant, Evergreen Nylon Recycling LLC has opened in Augusta, Ga., it is announced. The 85 million US dollar facility will process more than 200 million pounds of post-consumer nylon-6 carpet waste, representing nearly 20% of all discarded nylon-6 carpet. The company, a joint venture between DSM Chemicals North America and AlliedSignal Inc. Performance Polymers, uses a patented technology to convert nylon carpet into caprolactam, and will produce over 100 million pounds of it each year by an energy efficient closed loop system. The Evergreen system will allow AlliedSignal Performance Polymers to produce its Infinity Forever Renewable Nylon, for all applications of nylon 6, including carpet fibre.

EVERGREEN NYLON RECYCLING LLC

USA

Accession no.759561

Item 80

International Composites Expo '99. Conference proceedings.

Cincinnati, Oh., 10th-12th May 1999, session 11-D

FRP RECYCLING IN JAPAN
Nomaguchi K
(SPI,Composites Institute)

Alternative processes for the recycling of fibre-reinforced plastic (FRP), and their application in Japan, are briefly reviewed. Pulverised waste has been used in plastic mouldings for automotive applications, and in cement roof tiles. FRP may be burned in incinerators and used to heat water, or as an additive to cement kilns, where the resin acts as a fuel and the glass and filler become cement raw materials. Pyrolysis, in conjunction with metal catalysts, has been used to reduce the waste to oils or gases, and treatment with steam or supercritical water has also been successfully applied. 26 refs.

JAPAN

Accession no.759504

Item 81
International Composites Expo '99. Conference proceedings.
Cincinnati, Oh., 10th-12th May 1999, session 11-C
ENERGY BALANCE STUDY OF A NEWLY DEVELOPED RECYCLING SYSTEM FOR WASTE PLASTICS
Nomaguchi K; Hayashi S
Japan,Ship Research Institute
(SPI,Composites Institute)

A system for recycling fibre-reinforced plastics (FRP) is described. Superheated water is created using the heat from the combustion of waste plastics. This water is used to decompose FRP waste, the soluble product being available for recycling, and the solid residue which is mainly glass fibre is melted at 1400-1500 C and moulded into new products. Surplus water is used to produce steam for electricity generation. It was concluded that the process offered good economics, with acceptable energy efficiency whilst generating no pollution. 4 refs.

JAPAN

Accession no.759503

Item 82
Informations Chimie
No.411, Sept.1999, p.128-9
French
TREATMENT OF WASTES: ELIMINATION OF CHLORINE IN WASTE THERMOLYSIS
Fontana A; Laurent P; Jung C G; Gehrmann J; Beckmann M
Universite Libre de Bruxelles; Clausthaler Umwelttechnik-Institut GmbH

Results are presented of studies of the thermolysis of PVC-containing wastes in which calcium carbonate, calcium hydroxide, sodium carbonate and sodium hydrogen carbonate were used for the capture of chlorine. 2 refs.

BELGIUM; EUROPEAN COMMUNITY; EUROPEAN UNION; GERMANY; WESTERN EUROPE

Accession no.758765

Item 83
Polymer Journal (Japan)
31, No.9, 1999, p.714-6
DECOMPOSITION OF POLYETHYLENE 2,6-NAPHTHALENE DICARBOXYLATE TO CONSTITUENT MONOMERS USING SUPERCRITICAL METHANOL
Sako T; Sugeta T; Otake K; Yoda S; Takebayashi Y; Okajima I
Tsukuba,National Institute of Materials & Chemical Research

The depolymerisation of polyethylene 2,6-naphthalene dicarboxylate (PEN) to its constituent monomers and oligomer using supercritical methanol without a catalyst was investigated as a means of chemical recycling of waste PEN. Attention focused on the analysis of the decomposition products, which included both solids and liquids. The amount of gases produced was negligible, as the decomposition temperature was only 623 K. The solid phase consisted of unreacted PEN, undissolved monomer dimethyl 2,6-naphthalene dicarboxylate (DMN) and oligomer. The oligomer was defined as a product which was hydrolysed with sodium hydroxide aqueous solution to disodium 2,6-naphthalene dicarboxylate and ethylene glycol. The liquid phase contained ethylene glycol monomer, DMN monomer and oligomer. 5 refs.

JAPAN

Accession no.758005

Item 84
Polymers for Advanced Technologies
10, No.10, Oct.1999, p.596-602
HYDROTHERMAL RECYCLING OF PVC IN A BUBBLING FLUIDIZED BED REACTOR: THE INFLUENCE OF BED MATERIAL AND TEMPERATURE
Slapak M J P; van Kasteren J M N; Drinkenburg B A A H
Eindhoven,University of Technology

Hydrothermal recycling of poly(vinyl chloride) (PVC) to hydrogen chloride, a mixture of hydrocarbons and syngas, was investigated using a bench-scale bubbling fluidised bed reactor. The use of a quartz sand bed at 1150 K proved to be unsuitable, as conversion of PVC to gas was only about 25%, and large amounts of tar and char were formed. The use of porous gamma-alumina powder instead of quartz improved the gaseous yield to 69%. This was attributed to the catalytic activity and the large specific area of the alumina powder. The gaseous fraction consisted of hydrogen chloride, hydrogen, carbon monoxide, carbon dioxide and methane. No chlorine-containing organic compounds were detected. The gas yield was not improved by increasing the residence time, but was increased to 98% by increasing the reactor temperature to 1250 K. 12 refs.

EUROPEAN COMMUNITY; EUROPEAN UNION; NETHERLANDS; WESTERN EUROPE

Accession no.757098

Item 85

Polymer Recycling

4, No.1, 1999, p.41-55

RECOVERY OF WASTE ELECTRICAL AND ELECTRONIC EQUIPMENT

Ramlow G; Christill M

BASF AG

Electrical and electronic waste can be successfully recovered by pyrolysis following a suitable amount of separation. It has been demonstrated that by selecting the correct processing parameters, any halogenated hydrocarbons present are satisfactorily destroyed. Oil and gas from the pyrolysis are best used to heat the pyrolysis kiln. The metal containing pyrolysis coke can be treated in a secondary copper smelter without need for further processing.

EUROPEAN COMMUNITY; EUROPEAN UNION; GERMANY; WESTERN EUROPE

Accession no.756663

Item 86

Polyurethanes Expo '99. Conference proceedings. Orlando, Fl., 12th-15th Sept.1999, p.377-82

NEW ERA OF POLYURETHANE RECYCLING - FASCIA TO ROOF RAIL: SUSTAINABLE RECYCLING IN AUTOMOTIVE APPLICATIONS

You K K; Chang L P; Witte E A; Prokopyshen M H

BASF Corp.; DaimlerChrysler Corp.

(American Plastics Council,Alliance for the Polyurethanes Industry)

Due to the increasing need for finding alternative solutions to the disposal of PU waste, BASF has carried out extensive research in the field of chemical recycling of PUs, leading to a patented process for glycolysis of solid PU parts. In 1997, a PU glycolysis plant was started up at Philip Environmental Services in Detroit, Michigan utilising the BASF technology. This process is a 'one-pot' reaction requiring no separation and no by-product disposal. The plant is capable of generating up to 10 million lb of glycolysate per year. Currently two recycled polyols have been successfully commercialised: Pluracol RP2001 and Pluracol RP1464 polyols. These polyols are glycolised products of reaction injection moulding scrap. These two recycled polyols have a relatively high hydroxyl number and are suitable for flexible, semi-rigid and rigid PU applications. In some instances the glycolysate can be formulated into a system with up to 100% loading and still maintain the required physical performance. BASF and DaimlerChrysler have jointly developed a sustainable recycling application, the first commercialised process in the North American Free Trade Agreement region, that takes automotive-derived scrap to produce new automotive parts - discarded fascia become new roof rail. The properties and performance of these two recycled polyols are examined. The significance of the sustainable recycling application, a new beginning in PU chemical recycling, is discussed. It is shown that the finished PU parts containing glycolysates maintain all performance requirements. 6 refs.

USA

Accession no.755705

Item 87

Plastiques Modernes et Elastomeres

51, No.6, Aug./Sept.1999, p.14-7

French

PETP: IS CHEMICAL RECYCLING THE WAY FORWARD FOR THE 21ST CENTURY?

Renaudat E

The technical advantages of the chemical recycling of PETP bottles are discussed, and some developments in depolymerisation processes are examined. Particular attention is paid to glycolysis, hydrolysis and solvolysis processes respectively developed by TBI, Tredi and Eastman Chemical.

VALORPLAST SA; PETCORE; ADEME; ELF ATOCHEM SA; ENICHEM SPA; DSM NV; FINA CHEMICALS; BP CHEMICALS LTD.; TBI; EASTMAN CHEMICAL CO.; TREDI; CSTB; ECOLE CENTRALE DE PARIS

BELGIUM; EUROPEAN COMMUNITY; EUROPEAN UNION; FRANCE; ITALY; NETHERLANDS; UK; USA; WESTERN EUROPE

Accession no.752780

Item 88

Resources, Conservation & Recycling

23, Nos.1-2, 1998, p.47-56

ORGANIC SOLVENT EFFECTS ON WASTE PLASTICS-LIGNITE COLIQUEFACTION

Gimouhopoulos K; Doulia D; Vlyssides A; Georgiou D

Athens,National Technical University; Toronto,University

Converting the organic fraction of municipal solid waste into useful products, e.g. gas and liquid fuels, seems to be an option of great interest both environmentally and economically. This paper examines the results of co-liquefaction of low-grade coal, lignite, with post-consumer plastics. Special catalysts were prepared for this purpose and tested along with different types of organic solvents. The presence of these solvents during the co-liquefaction process almost doubled total solids conversion into gas and liquid products. Decane and toluene were found to be the best organic solvents for co-liquefaction of lignite with HDPE and PS, respectively.

Total solids conversion reached almost 90% when a two-stage process was employed. 15 refs.

CANADA; EUROPEAN COMMUNITY; EUROPEAN UNION; GREECE; WESTERN EUROPE

Accession no.752715

Item 89

Composites-French/English

No.35, Sept/Oct.1999, p.58-61

French; English

FRP RECYCLING IN JAPAN

Japan's commitment to FRP recycling is discussed. As early as 1974 the Japan Reinforced Plastics Society established the Research Committee for FRP Waste Treatment. Details are given of more recent associations and organisations, set up to deal with recycling issues. FRP products in Japan are listed by application, with details of production in tons. Estimated total FRP waste volume is put at 282,000 tons for 1998. Recycling techniques used are examined and include thermal recycling, pyrolysis, chemical recycling, and material recycling.

JAPAN

Accession no.751604

Item 90

Polymer International

48, No.9, Sept.1999, p.885-8

KINETICS OF GLYCOLYSIS OF POLYETHYLENE TEREPHTHALATE WITH ZINC CATALYST

Jong-Wu Chen; Leo-Wang Chen; Wu-Hsun Cheng

Taiwan,National University; Taiwan,Cheng Gung University

The glycolysis of PETP melts with ethylene glycol was examined in a pressurised reactor, glycolysis being used for depolymerisation of PETP in recycling. The kinetics of the glycolysis reaction were studied. The rate constants for glycolysis without addition of catalyst were calculated at four different temps., yielding an activation energy of 108 kJ/mol. In comparison, the rate constants for glycolysis with addition of zinc acetate were also calculated at four different temps., yielding an activation energy of 85 kJ/mol. It was found that the activation energy of glycolysis with addition of zinc acetate was lower than that of glycolysis without addition of catalyst. Zinc acetate thus had a catalytic effect on PETP glycolysis at temps. between 235 and 275C. The effect of catalyst concentration on reaction rate constants was also examined. Below a critical catalyst concentration, the rate constant for glycolysis was linearly dependent on catalyst concentration. 14 refs.

TAIWAN

Accession no.751071

Item 91

ENDS Report

No.296, Sept.1999, p.19

PALLET PLANT MAY OFFER MARKET FOR PLASTIC PACKAGING WASTE

Efforts to increase the recycling rate of plastics packaging are discussed with reference to the initiative by Quarrtik to build a pallet manufacturing plant which would use up to 25,000 tonnes per year of post-consumer mixed plastic packaging. Valpak, the largest of the compliance schemes has providing support to the company in securing a site and supplies of waste plastics for reprocessing. If the project goes ahead, it will renew the competition between wood and plastic pallet suppliers, and the plant would double the number of plastic pallets in circulation to around 2 million. Meanwhile, BP Amoco's plans for a feedstock recycling plant have suffered due to EC proposals to revise the EC packaging Directive, suggesting that feedstock recycling of plastics should be relegated to the status of energy recovery if the material produced was fed into a refinery to manufacture fuels, and the amounts recycled would not contribute to the 16% minimum recycling rate for each of the main packaging materials.

BP AMOCO; QUARRTIK

EUROPEAN COMMUNITY; EUROPEAN UNION; UK; WESTERN EUROPE

Accession no.750414

Item 92

Resources, Conservation & Recycling

23, No.3, 1998, p.163-81

CATALYTIC PLASTICS CRACKING FOR RECOVERY OF GASOLINE-RANGE HYDROCARBONS FROM MUNICIPAL PLASTIC WASTES

Buekens A G; Huang H

Brussels,Free University

Developments in plastics cracking, a process developed for recycling of plastics waste into useful petrochemical materials, are reviewed. It is shown that, under thermal cracking conditions, plastics wastes can be decomposed into three fractions, i.e. gas, liquid and solid residue. The liquid products are usually composed of higher boiling point hydrocarbons. By use of fluid cracking catalysts and reforming catalysts, more aromatics and naphthenes in the C6-C8 range can be produced, which are valuable gasoline-range hydrocarbons. Industrial aspects of implementing this technology are considered. 47 refs.

BELGIUM; EUROPEAN COMMUNITY; EUROPEAN UNION; WESTERN EUROPE

Accession no.749062

Item 93

European Chemical News

71, No.1876, 20th-26th Sept.1999, p.47

RESCUING RECYCLING
Johnston S

Veba Oel's announcement that it will shut its 80,000 tonne/year plastics-to-petrochemicals plant at Bottrop, Germany is discussed with reference to European recycling efforts. The PVC industry has recently announced its investment of 3 million US dollars in a 2-3 year project to develop a feedstock recycling technology. The project will be funded by the European Council for Vinyl Manufacturers. PVC manufacturers represented by ECVM have a huge vested interest in proving that PVC feedstock recycling is possible, because if not, the sustainability of the world's second largest commodity plastic is said to be in question. Trials of the five technologies were carried out, and a gasification technology from Linde-KCA was chosen, brief details of which are given.

VEBA OEL
EUROPEAN COMMUNITY; EUROPEAN UNION; FRANCE; GERMANY; WESTERN EUROPE

Accession no.748978

Item 94
PVC '99. Conference proceedings.
Brighton, 20th-22nd April 1999, p.123-31
OPTIONS FOR PVC FEEDSTOCK RECYCLING RESULTS OF THE ECVM R&D PROJECT ON PVC FEEDSTOCK RECYCLING PROCESSES
Buehl R
EVC International SA
(IOM Communications Ltd.; BPF)

Feedstock recycling processes for treating PVC rich waste streams should be capable of recovering both the chlorine and hydrocarbon contents. ECVM sponsored research and development work to identify potential technologies for building a pilot plant is described. 3 refs.

BELGIUM; EUROPEAN COMMUNITY; EUROPEAN UNION; WESTERN EUROPE

Accession no.747883

Item 95
Warmer Bulletin
No.68, Sept. 1999, p.22-3
WASTE PLASTICS CO-COMBUSTION WITH MSW

The results of studies are discussed into the co-combustion of plastics with respect to the recycling of scrap plastics from the building, packaging, automotive, electrical and electronic industries. Tests were carried out by APME on plastic waste from these end-use industries in a typical large scale EfW facility with respect to operational and environmental problems.

APME
WESTERN EUROPE

Accession no.747549

Item 96
Chimica e l'Industria
81, No.3, April 1999, p.355-60
Italian
POST-CONSUMER PETP: FROM PROBLEM TO OPPORTUNITY
Chiacchio G; Malinconico M; Santacesaria E; Di Serio M
Istituto di Ricerca e Tecnologia delle Materie Plastiche; Napoli,Universita Federico II

Methods used in the recycling of post-consumer PETP bottles are examined, including mechanical and chemical recycling and incineration with energy recovery. Some new developments aimed at overcoming problems associated with these processes and improving the properties of recycled materials are discussed, and reference is made to techniques for producing phthalate plasticisers for thermoplastics and monomers for unsaturated polyesters through the chemical recycling of PETP. 16 refs.

REPLASTIC
EU; EUROPEAN COMMUNITY; EUROPEAN UNION; ITALY; WESTERN EUROPE; WESTERN EUROPE-GENERAL

Accession no.745818

Item 97
Plastics News(USA)
11, No.21, 12th July 1999, p.1/35
EASTMAN TACKLES PETP RECYCLING
Toloken S

Eastman Chemical is starting a pilot depolymerisation plant that it hopes can provide a cost-effective solution for some new hard-to-recycle PETP bottles. In the laboratory, the process has been able to handle all the different coloured PETP and all the barrier layers that have been tested. The process produces food-grade material.

EASTMAN CHEMICAL CO.
USA

Accession no.742848

Item 98
Additives for Polymers
July 1999, p.9
CLOSED-LOOP RECYCLING PROCESS FOR PVC IS BASED ON SOLVENTS

New recycling technology using solvents has proved effective for all types of PVC compounds tested so far: cables, pharmaceutical blister packs, floor covering and automotive dashboards. The process, developed by Solvay, is called Vinyloop. Its first industrial application is now being developed and is due to be commercialised in 2001. The process was developed in response to a request for help in recycling from the PVC and polyester tarpaulin and canvas manufacturer, Ferrari Textiles Techniques. Seeing PVC as an ideal base for many compound products offering high technical properties, it was appreciated that the compound could not be recycled unless the components could be separated. Details are given.

SOLVAY & CIE.SA

EUROPEAN COMMUNITY; EUROPEAN UNION; FRANCE; WESTERN EUROPE

Accession no.741722

Item 99

Polymer Plastics Technology and Engineering

38, No.3, 1999, p.471-84

RECYCLING NYLON 6 CARPET TO CAPROLACTAM

Braun M; Levy A B; Sifniades S

AlliedSignal Inc.

A process for the depolymerisation of Nylon 6 carpet fibre in the presence of steam under medium pressure (800 to 1500 KpA, 100 to 200 psig) is described. The feasibility of the scheme was demonstrated using a small laboratory apparatus and the best run produced a 95% yield of crude caprolactam. The data obtained were used to construct a computer model of the process for both batch and continuous flow stirred reactors. 6 refs.

USA

Accession no.741249

Item 100

Polymer Plastics Technology and Engineering

38, No.3, 1999, p.459-70

NYLON 66, NYLON 46, AND PET PHASE-TRANSFER-CATALYSED ALKALINE DEPOLYMERISATION AT ATMOSPHERIC PRESSURE

Polk M B; Leboeuf L L; Shah M; Won C-Y; Hu X; Ding W

Georgia,Institute of Technology

A method for the depolymerisation of PETP fibres using quarternary ammonium salt phase transfer catalysts in saponification processes at atmospheric pressure and temperatures as low as room temperature is reported. Terephthalic acid was produced in yields as high as 93%. Also reported are similar processes for the depolymerisation of nylon 66 and nylon 46 fibres. Nylon 46 oligomers produced were repolymerised using solid-state polymerisation to produce high molecular weight nylon 46. Nylon 66 was depolymerised to produce oligomers and adipic acid in reasonable yields. 11 refs.

USA

Accession no.741248

Item 101

Carl Hanser, Munich, 1996, pp.893. 135.00. 8(13)

RECYCLING AND RECOVERY OF PLASTICS

Brandrup J; Bittner M; Michaeli W; Menges G (IKV)

Recycling and recovery of plastics went through a tremendous development during the last years, due mainly to a corresponding legal framework. This book describes all aspects of this development: it cites legal requirements for recycling, it provides details on life cycle analysis, it covers technical and other aspects of sorting, pretreatment, mechanical and feedstock recycling as well as of energy recovery. Markets for recycled plastics and economical aspects are discussed as well as developments in Europe, Japan and the USA. It is demonstrated that plastics are recyclable - recyclable in as many individual ways as there are for their application. This book is a necessary for everyone concerned with plastics.

EUROPEAN COMMUNITY; EUROPEAN UNION; GERMANY; WESTERN EUROPE

Accession no.739325

Item 102

Japan Chemical Week

40, No.2031, 1st July 1999, p.9

CONTAINERS RECYCLING LAW TO BE ENFORCED NEXT APRIL

Under the Japanese Law for Promotion of Sorted Collection and Recycling of Containers and Packaging, it will become compulsory from next April to collect plastic containers and wrappings, excluding PETP bottles, paper containers and wrapping separately, and to recycle them. For plastic containers and wrappings, in particular, it was decided to employ a new recycling method to process waste plastics into chemical materials in coke ovens, in addition to the four existing methods. These are material recycling, conversion to oil, blast furnace reducing agents and gasification.

JAPAN

Accession no.739149

Item 103

Journal of Applied Polymer Science

73, No.7, 15th Aug.1999, p.1139-43

MECHANISM STUDIES ON THE CATALYTIC DEGRADATION OF WASTE POLYSTYRENE INTO STYRENE IN THE PRESENCE OF METAL POWDERS

Xi Guoxi; Liang Rui; Tang Qinhu; Li Jinghua

Henan,University

The effects of aluminium, zinc, iron, nickel and copper powders on the thermal degradation of waste PS were studied. The results showed that the catalytic effects of metal powders were related to their activities. The catalytic effects increased with increasing activities of metals. It was suggested that PS degraded through a transient intermediate in the presence of metal powders and that the degradation of the transient intermediate was the rate-determining step. 10 refs.

CHINA

Accession no.739021

Item 104

ACS, Polymeric Materials Science and Engineering. Vol.76. Conference proceedings.
San Francisco, Ca., Spring 1997, p.578-9. 012

CATALYTIC CONVERSION PROCESS FOR RECYCLING NAVY SHIPBOARD PLASTIC WASTES

Allred R E; Doak T J; Busselle L D; Gordon B W; Harrah L A; Hovt A E
Adherent Technologies Inc.
(ACS,Div.of Polymeric Materials Science & Engng.)

The plastics recycling industry recognises four categories or types of recycling for scrap or waste plastics: primary - conversion into products equivalent to the original material; secondary - conversion into products with reduced performance; tertiary - producing chemicals or fuels; and quaternary - recovering energy through incineration. Of these major waste reduction methods, only tertiary recycling or quaternary recycling methods allow for the efficient treatment of waste streams containing complex mixtures of plastics, metals and inorganics. Adherent Technologies is currently investigating tertiary recycling or catalytic depolymerisation methods in which low heat and catalysis are utilised to generate mixtures of low molecular weight hydrocarbon materials as products. The mixture of products depends largely upon the feedstock; more complex feedstock will generate a more complex mixture of depolymerisation products. Materials currently being successfully recycled using this technology include tyres, fibre-reinforced composites, printed circuit boards and computer casing materials. The tertiary recycling of Navy Shipboard Plastic Waste Processor (SPWP) product using catalytic depolymerisation is described. Shipboard plastic waste consists largely of packaging materials and food service items. In addition to its complexity, this waste is extremely variable in content and heavily contaminated with food residue, paper products, and metals. An analysis of process feasibility, including reproducibility, product analysis and economics are presented. 2 refs.

CANADA

Accession no.738628

Item 105

Plastics News(USA)
11, No.14, 24th May 1999, p.20

DUPONT KICKING OFF PILOT RECYCLING PLANT

Pryweller J

It is reported here that DuPont is planning to launch a pilot plant in 2000, to test a new, patented process for chemically recycling nylon 6/6 resin. Full details are given of the process, called ammonolysis.

DUPONT AUTOMOTIVE; ALLIEDSIGNAL INC.; DSM CHEMICALS NORTH AMERICA INC.; EVERGREEN RECYCLING LLC; VISTEON AUTOMOTIVE SYSTEMS; WELLMAN INC.

CANADA; USA

Accession no.737719

Item 106

Patent Number: US 5886057 A 19990323

PRODUCTION OF DICARBOXYLIC ACIDS

Harvie J L; Heppell S M
DuPont de Nemours E.I.,& Co.

Dicarboxylic acids or esters thereof are recovered from solid phase polyester materials, such as post-consumer products and factory scrap, by subjecting the polyester to at least two hydrolysis stages in at least the first of which the amount of water used is substantially less than needed to effect total conversion of the polyester to the dicarboxylic acid. Also the diol content is controlled in the course of carrying out the hydrolysis. The hydrolysis reactions may be preceded by reaction of the polyester with a diol, the resulting depolymerisation products then being hydrolysed.

USA

Accession no.737440

Item 107

Polymer Recycling
3, No.3, 1997/98, p.227-37

UNSATURATED POLYESTER RESINS ON THE BASE OF CHEMICAL DEGRADATION PRODUCTS OF PET WASTE FOR VARNISHES

Viksne A; Rence L; Berzina R; Kalnis M
Riga,Technical University

A description is given of a comparative study of the glycolysis of PETP waste soft drinks bottles by various mixtures of EG and DEG with subsequent polyesterification of the glycolysed products by maleic anhydride in order to obtain unsaturated polyesters suitable for the production of varnishes. The processing characteristics such as viscosity, exotherm temperatures of curing, compatibility of resins with monomers was investigated with respect to the type and amount of reactive monomers. The mechanical properties of varnishes produced were analysed. 13 refs.

LATVIA

Accession no.736641

Item 108

Polymer Recycling
3, No.3, 1997/98, p.173-80

UNSATURATED POLYESTER RESINS FROM POLY(ETHYLENE TEREPHTHALATE) WASTE: SYNTHESIS AND CHARACTERISATION

Abdel-Azim A A; Mekewi M A; Gouda S R
Egyptian Petroleum Research Institute; Ain Shams,University; Egypt,Military Technical College

Different glycolysed oligomers were prepared by depolymerisation of PETP waste in the presence of manganese acetate catalyst. Diethylene glycol, triethylene glycol, propylene glycol and mixtures thereof were used for glycolysis. The hydroxyl value of the glycolysed products before and after removing the free glycol were determined, and the amount of free glycol in each case was analysed. These glycolysed products were reacted with maleic anhydride to prepare a series of unsaturated polyesters having different molecular weights. These were then dissolved in styrene monomer and their curing behaviour was investigated and compared with the curing behaviour of unsaturated polyester made from virgin materials. The effect of accelerator and initiator concentrations on the curing characteristics of the recycled resins was studied, and the mechanical properties and hardness were measured and correlated to their molecular structure. 21 refs.

EGYPT

Accession no.736636

Item 109

Chemical Engineering

106, No.6, June 1999, p.54/8

DISCARDED CARPETING YIELDS VALUABLE, REUSABLE FEEDSTOCKS

Crabb C

This fall, the closed-loop Evergreen Nylon Recycling plant will start up in the US, a joint venture of DSM Chemicals North America and AlliedSignal. The facility will recover 45,000 m.t./year of caprolactam by depolymerising the fibres from 100,000 m.t./year of discarded nylon-6 carpets. Meanwhile in Germany, Lurgi is building the Polyamid 2000 AG facility. It will process 120,000 m.t./year of carpet waste and recover 10,000 m.t./year of caprolactam from nylon-6 carpets and 13,000 m.t./year of nylon-6-6 from nylon-6-6 carpets.

EVERGREEN NYLON RECYCLING LLC; LURGI

EUROPEAN COMMUNITY; EUROPEAN UNION; GERMANY; USA; WESTERN EUROPE

Accession no.736166

Item 110

Polimery Tworzywa Wielkoczasteczkowe

44, No.1, 1999, p.1-5

Polish

CHEMICAL RECYCLING OF WASTE SATURATED POLYESTERS AND URETHANE POLYMERS TO YIELD RAW MATERIALS FOR THE PRODUCTION OF POLYURETHANES

Kasperski M; Spychaj T

Szczecin,Polytechnic

Techniques for chemical recycling of waste saturated polyesters, mainly poly(ethylene terephthalate) and of waste polyurethanes to produce polyurethanes are reviewed. 29 refs. Articles from this journal can be requested for translation by subscribers to the Rapra produced International Polymer Science and Technology.

EASTERN EUROPE; POLAND

Accession no.735069

Item 111

High Performance Plastics

June 1999, p.6-7

NEW ROUTE TO ENGINEERING PLASTICS: PRODUCED FROM WASTE CARPETS

A three-year joint European project, RECAM, recommends that it should be possible to collect more than 50% of carpet waste in Western Europe. High-grade materials such as PA and PP could be recovered for the manufacture of engineering plastics compounds and more than 8 million Gigajoules of energy could be recovered from the remainder. At the heart of the project are chemical recycling processes developed by both DSM and Enichem.

DSM NV

EUROPEAN COMMUNITY; EUROPEAN UNION; NETHERLANDS; WESTERN EUROPE

Accession no.734187

Item 112

Patent Number: US 5869654 A 19990209

PROCESS FOR DEPOLYMERISING POLYCAPROLACTAM PROCESSING WASTE TO FORM CAPROLACTAM

Sifniades S; Levy A B

AlliedSignal Inc.

The polycaprolactam waste is contacted with superheated steam in the absence of added catalyst at a temperature of about 250 to 400C and a pressure in the range of about 1.5 to 100 atm. and substantially less than the saturated vapour pressure of water at the temperature at which a caprolactam-containing vapour stream is formed. The resulting caprolactam may then be used in the production of engineered resins and fibres.

USA

Accession no.729580

Item 113

Popular Plastics and Packaging

44, No.3, March 1999, p.76-86

PROGRESS IN RECYCLING OF POLYURETHANES

Frisch K C

Detroit,Mercy University

Recycling of waste materials was classified in general and recycling technologies of PUs were described based on literature data. In particular, physical and chemical recycling methods of PU were described; energy recovery was mentioned briefly. Physical recycling methods include separation, granulation, densification and/or disintegration operations, then reusing powdered material for PU production. These methods are applicable for flexible and rigid PU foams, as well as for reaction injection moulded products. Amongst chemical recycling methods considered were pyrolysis, hydrolysis, glycolysis, recovery with alkanolamines, and petrochemical feedstock processing. In particular, glycolysis was described and mechanisms of chemical reactions were presented. 26 refs.

USA

Accession no.729287

Item 114

Kunststoffe Plast Europe

89, No.2, Feb. 1999, p.25-7. (Translated from Kunststoffe 89 (1999), 2, pp.85-8

RECYCLING OF PUR FLEXIBLE FOAMS

Bauer G; Kugler M; Chakrabarti R S

Aalen,Fachhochschule; Regra Recycling GmbH; Elastogran GmbH

The use of a solvolysis chemical process for the recycling of flexible polyurethane foam into liquid polyol feedstocks is discussed. The partial chemical degradation of polyurethane is carried out with small amounts of carboxylic acids. The oligomeric degradation products produced during the process are stabilised by steric stabilisation so they can be homogeneously mixed with polyetherols. Details are given of product properties, process costs, effect on foaming behaviour, mechanical properties, and application properties. 7 refs.

EUROPEAN COMMUNITY; EUROPEAN UNION; GERMANY; WESTERN EUROPE

Accession no.726067

Item 115

Brussels, APME, 1995, pp.40. 29cms. 8(13)

LIFE CYCLE ANALYSIS OF RECYCLING AND RECOVERY OF HOUSEHOLDS PLASTICS WASTE PACKAGING MATERIALS. SUMMARY REPORT

APME

This summary report covers life cycle analysis of recycling and recovery of households plastics waste packaging materials. Main sections include recycling and recovery methods investigated, procedure and results. Arising from the German law relating to the avoidance and disposal of wastes this report examines mechanical recycling, film recycling, feedstock recycling and energy recovery.

Accession no.725346

Item 116

Urethanes Technology

16, No.1, Feb./March 1999, p.6

FRENCH FIRM ADDS APPS

Groupe TBI is investing about FFr3m to build a 15,000 t/y aromatic polyester polyols plant on a greenfield site at Issoire, near Lyons. The novel, patented process uses post-consumer waste bottles made from PETP as feedstock. PU and modified PIR foams made using the APPs have excellent fire performance and good dimensional stability.

GROUPE TBI

EUROPEAN COMMUNITY; EUROPEAN UNION; FRANCE; WESTERN EUROPE

Accession no.718653

Item 117

Plastics Southern Africa

28, No.5, Nov.1998, p.18/22

BACKGROUND TO MICROWAVE PROCESS FOR RECOVERY OF PMMA WASTE

Researchers at AECI's Research and development Department have developed a novel microwave depolymerisation process for the thermal decomposition of polymethyl methacrylate and the recovery of the monomer methyl methacrylate. This comprehensive article supplies a detailed explanation and examination of the process which has been patented in South Africa. The microwave technology provides a purer product which will simplify downstream processing and reduce effluent volume and chemical consumption.

AECI

SOUTH AFRICA

Accession no.718276

Item 118

Plastics and Rubber Weekly

No.1776, 5th March 1999, p.10

INSULATION PANELS TO ABSORB BOTTLE WASTE

Over 4 billion PETP bottles will be available for collection across Europe in 1999. PUR Products has introduced technology into the UK which involves glycolysis of post-consumer PETP into materials for the manufacture of rigid urethane foams for building insulation. This application offers a substantial new market for aromatic polyester polyols derived from glycolised PETP recyclate.

PUR(PRODUCTS)LTD.

EUROPEAN COMMUNITY; EUROPEAN UNION; UK; WESTERN EUROPE

Accession no.717127

Item 119

Plastverarbeiter

46, No.1, Jan.1995, p.20-5

German

RECYCLING AND UTILISATION OF POLYURETHANES - POSSIBILITIES AND LIMITS, PART I

Weigand E

First of all the author explains every potential for the application, use and recycling of polyurethanes (PUR). The focal point of this first part of a multipartite review contribution covers the various possibilities for utilising PURs. Their use is discussed as raw material recycling in glycolysis, hydrolysis, pyrolysis, hydrogenation and gas production. They are reviewed as material in particle bonding, powder bonding, injection moulding and extrusion. Their application is also considered as a source of energy for burning household rubbish, kilns, low burning or smouldering processes. The potential of tailor-made PURs is discussed for many applications. There is a statistical analysis of the range of applications for PURs in Western Europe according to use and service life. Such applications covered include building work, packaging, leisure and clothing, domestic appliances, tank insulation, vehicles, furniture and mattresses. Also quantified according to selected properties are widely used PURs such as glass fibre-reinforced PURs, elastomer thermoplastic PURs, soft and hard integral foams and cellular foams.

SUEDDEUTSCHES KUNSTSTOFF-ZENTRUM

EUROPEAN COMMUNITY; EUROPEAN UNION; GERMANY; WESTERN EUROPE

Accession no.716577

Item 120

Macromolecular Symposia
Vol.135, Dec.1998, p.295-314

POLYMER RECYCLING: THERMODYNAMICS AND ECONOMICS

Stein R S
Massachusetts,University

Thermodynamic and economic aspects of polymer recycling are considered. The objective of recycling is two-fold, the reduction of rubbish and the burden of waste polymers on the environment and the conservation of resources. It is claimed that the environmental burden should be considered as part of the cost for employing polymers. The reduction of polymer input to landfills by source reduction, materials substitution, recycling, incineration and/or degradation or reduction to low molec.wt. compounds is discussed. It is shown that, while the value of separated polymers is significantly greater than that of commingled ones, the separation is often difficult and requires an energy investment for 'demixing'. With present technology, this is not justified for more than about 25% of the polymer waste feedstock. For the rest, energy recovery through incineration appears to be the thermodynamically and economically sensible route. 25 refs. (IUPAC Working Party on Recycling of Polymers, Prague, July 1997)

USA

Accession no.715506

Item 121

Macromolecular Symposia
Vol.135, Dec.1998, p.237-45

CHLORINATED PRODUCTS OF PLASTIC PYROLYSIS

Blazso M
Hungarian Academy of Sciences

The formation of various chlorinated products in pyrolysis of polymers and plastics additives was studied. The formation of chlorobenzenes (in addition to the monomers) from polychlorostyrene and polyvinylbenzyl chloride was observed. Hydrogen chloride was only produced from these polymers at above 600C when the chlorine atoms were cleaved off and abstracted hydrogen. A similar process took place in aromatic chlorine-containing dyes, in which the strong aromatic molecular structure prevented the thermal cleavage of chloroaromatic volatile products. It was observed that cupric and ferric chlorides chlorinated phenolic thermal decomposition products of plastics materials which originated either from the polymer or from the stabiliser. The highest yields of chlorophenols were obtained in pyrolysis at around 700C. 15 refs. (IUPAC, 38th Microsymposium on Recycling of Polymers, Prague, July 1997)

EASTERN EUROPE; HUNGARY

Accession no.715502

Item 122

Macromolecular Symposia
Vol.135, Dec.1998, p.205-14

CONVERSION OF POLYMERS AND BIOMASS TO CHEMICAL INTERMEDIATES WITH SUPERCRITICAL WATER

Arai K
Tohoku,University

Results are reported of recent studies on the conversion of polymers and biomass to chemical intermediates and monomers by using subcritical and supercritical water as the reaction solvent. The reactions of cellulose in supercritical water are shown to be rapid and to proceed to 100% conversion with no char formation, these reactions showing a significant increase in hydrolysis products and lower pyrolysis products when compared with reactions in subcritical water. There is also a jump in the reaction rate of cellulose at the critical temp. of water. If the methods used for cellulose are applied to synthetic polymers, such as PETP or polyamide, high liquid yields can be achieved although the reactions require about 10 min for complete conversion. The reason for this is the heterogeneous nature of the reaction system. For PE, higher yields of short-chain hydrocarbons, higher alkene/alkane ratios and higher conversions are obtained in supercritical water than those obtained by pyrolysis. 18 refs. (IUPAC, 38th Microsymposium on Recycling of Polymers, Prague, July 1997)

JAPAN

Accession no.715499

Item 123
Macromolecular Symposia
Vol.135, Dec.1998, p.175-82
NEW CHEMICAL RECYCLING METHODOLOGIES: HYDROUS PYROLYSIS TO RECOVER MONOMERS FROM POLYOLEFINS
Audisio G; Bertini F; Beltrame P L; Bergamasco L; Castelli A
Istituto di Chimica delle Macromolecole del CNR; Milan,University

The thermal degradation of PS was investigated in the presence of water under subcritical conditions (hydrous pyrolysis). The experiments were carried out in closed systems under an inert atmosphere, in the temp. range 300-350C, at pressures up to 180 atm, for 1-120h. The results showed that the presence of water increased the yields of volatile products, mainly in the first steps of the pyrolytic process, with a high selectivity in the monomer. In order to improve the phase contact between water and polymer during the pyrolysis, some runs were carried out on silica-supported PS and using a stirred reactor. Under these conditions, the above effects appeared to be greatly enhanced. The best recovery of styrene, close to 71%, was achieved at 320C and 8h. 8 refs. (IUPAC, 38th Microsymposium on Recycling of Polymers, Prague, July 1997)

EUROPEAN COMMUNITY; EUROPEAN UNION; ITALY; WESTERN EUROPE

Accession no.715496

Item 124
Macromolecular Symposia
Vol.135, Dec.1998, p.137-45
NEW TRENDS IN CHEMICAL RECYCLING OF POLYETHYLENE TEREPHTHALATE
Spychaj T; Paszun D
Szczecin,Technical University

The chemical recycling of PETP by non-conventional transesterification and ammonolysis/aminolysis methods is discussed on the basis of data in the literature and the authors' own experimental data. The products obtained by deep PETP degradation using allylamine (N,N'-diallylterephthaldiamide), triethanolamine and other alkanolamines were prepared and characterised by DSC and elemental analysis. 26 refs. (IUPAC, 38th Microsymposium on Recycling of Polymers, Prague, July 1997)

EASTERN EUROPE; POLAND

Accession no.715492

Item 125
Macromolecular Symposia
Vol.135, Dec.1998, p.113-20
POLYMER CRACKING - NEW HYDROCARBONS FROM OLD PLASTICS
Hardman S; Wilson D C
BP Chemicals

European Union legislation sets demanding targets for the recycling of all materials, including plastics packaging materials. The progress made by an industry consortium, led by BP Chemicals, in developing technology to help meet the recycling targets is described. The use of the polymer cracking process as a method of returning mixed plastics waste to the mainstream hydrocarbon processing industry is discussed and the implementation of polymer cracking is considered. 2 refs. (IUPAC, 38th Microsymposium on Recycling of Polymers, Prague, July 1997)

EUROPEAN COMMUNITY; EUROPEAN UNION; UK; WESTERN EUROPE

Accession no.715489

Item 126
Macromolecular Symposia
Vol.135, Dec.1998, p.63-81
TECHNICAL METHODS IN PLASTICS PYROLYSIS
Buekens A G; Schoeters J G
Brussels,Free University; Groep T

A survey is presented of laboratory, pilot and industrial scale reactors, pyrolytic extruders, stirred tanks and single- or double-bed fluidised systems. The use of plastics in oil refinery-based processes, such as fluid catalytic cracking, hydrocracking, viscosity breaking and delayed coking, and also (coal-derived) hydrogenation and gasification, is briefly covered. The product distribution in a pyrolysis process is shown to have to be derived experimentally, with the operating temp., heating rate, catalytic effects, residence time of feed material and reaction products, reaction pressure, possible supply of reactive gases and, of course, the chemical composition and structure of the original plastics as major factors. Some potential operating problems which are being studied in a EU-funded project with the acronym Cycleplast, a joint project of five teams from Central and Western Europe, are discussed. 9 refs.(IUPAC, 38th Microsymposium on Recycling of Polymers, Prague, July 1997)

BELGIUM; EUROPEAN COMMUNITY; EUROPEAN UNION; WESTERN EUROPE

Accession no.715486

Item 127
Macromolecular Symposia
Vol.135, Dec.1998, p.35-41
STEPWISE PYROLYSIS FOR RECYCLING OF PLASTIC MIXTURES
Bockhorn H; Hornung A; Hornung U
Karlsruhe,University

For chemical recycling of plastics refuse, a cascade of cycled-spheres reactors was developed which combined separation and decomposition of polymer mixtures by stepwise pyrolysis at moderate temps. In low-temp. pyrolysis, mixtures of PVC, PS and PE or PS, polyamide-

6 and PE were separated into hydrogen chloride, styrene and polyamide-6 and aliphatic compounds from PE decomposition. Compared with the low-temp. pyrolysis of the single components, some interactions between the polymers were found when pyrolysing mixtures, Some mechanistic aspects of these interactions are discussed. 18 refs. (IUPAC, 38th Microsymposium on Recycling of Polymers, Prague, July 1997)

EUROPEAN COMMUNITY; EUROPEAN UNION; GERMANY; WESTERN EUROPE

Accession no.715483

Item 128

Gummi Fasern Kunststoffe

51, No.2, Feb.1998, p.126-30

German

CHEMICAL RECYCLING OF NON-WOVEN POLYAMIDE LINING WASTE

Meusel E; Seyfarth E; Taeger E

Thueringisches Institut fuer Textil- & Kunststoff-Forschung eV

Using the fact that polyamides are preferentially accessible to chemical attack at their molecular linkage points, waste materials from polyamide-based non-woven linings were degraded under mild reaction conditions to obtain low-molecular oligoamide diacids, which in a subsequent condensation process were combined with aliphatic diamines to obtain high-molecular copolyamides. These copolyamides can be used as melt adhesives in textile applications to obtain more compact non-woven linings and for bonding of textile linings. 8 refs. Articles from this journal can be requested for translation by subscribers to the Rapra produced International Polymer Science and Technology.

EUROPEAN COMMUNITY; EUROPEAN UNION; GERMANY; WESTERN EUROPE

Accession no.713790

Item 129

Materiaux & Techniques

96, Nos.11-12, Nov./Dec.1998, p.47

French

FIRST POLYAMIDE-6 FROM POST-CONSUMER RECYCLING

AlliedSignal's Infinity, Forever Renewable Nylon, prepared by recycling of polyamide-6, is briefly described. The polyamide is treated by depolymerisation, purification of the caprolactam monomer and repolymerisation. The new resin is said to exhibit the same properties as those of virgin polyamide-6.

ALLIEDSIGNAL

EUROPEAN COMMUNITY; EUROPEAN UNION; FRANCE; WESTERN EUROPE

Accession no.713469

Item 130

Kunststoffe Plast Europe

85, No.11, Nov. 1995, p.33-5

RECYCLING OF CONTAMINATED PU

Kettemann B U; Melchiorre M; Munzmay T; Rasshofer W

Daimler-Benz AG; Bayer AG

The recycling of contaminated polyurethane waste is described with reference to a glycolysis process developed by Bayer and Daimler-Benz in which the wastes from three-layer composite instrument panels are treated. The reuse of contaminated PU is achieved by dispersely integrating the impurities in the secondary polyol during the glycolytic dissociation. The impurities are modified in such a way that they act as a filler in the secondary polyol. 3 refs.

EUROPEAN COMMUNITY; EUROPEAN UNION; GERMANY; WESTERN EUROPE

Accession no.709871

Item 131

Urethanes Technology

15, No.6, Dec.1998/Jan.1999, p.32-3

COMMERCIAL EVALUATION OF RECYCLING PROCESS BEGINS IN JOINT-VENTURE UNIT

Reed D

ICI Polyurethanes and du Vergier are evaluating a PU recycling method. The three-year project aims to use a pilot plant to demonstrate the practicality of the split-phase glycolysis process that ICI has developed. Work will initially focus on flexible foams based on MDI and specially made at ICI's Rozenberg plant. In the second phase, the unit will use post-industrial waste. Assuming the trials are successful, a full-scale unit to handle at least 5000 t/y of scrap foam will be built.

ICI POLYURETHANES; DU VERGIER E.,& CO.LTD.

EUROPEAN COMMUNITY; EUROPEAN UNION; UK; WESTERN EUROPE

Accession no.709467

Item 132

Urethanes Technology

15, No.6, Dec.1998/Jan.1999, p.28/31

REGRA CLAIMS TO HAVE DEVELOPED AN ECONOMIC RECYCLING PROCESS

Knoedgen M

Regra Recycling has developed a chemical recycling process which, it claims, makes PU recycling a practical and economic option. The process has already been used with rigid and semi-rigid foams, SRIM and, recently, has been further developed for recycling flexible PU foams. The company sells complete recycling plants as well as offering commission recycling.

REGRA RECYCLING GMBH

EUROPEAN COMMUNITY; EUROPEAN UNION; GERMANY; WESTERN EUROPE

Accession no.709466

Item 133
Urethanes Technology
15, No.6, Dec.1998/Jan.1999, p.26-7
RECYCLING POLYURETHANES: LOTS OF WAYS TO DO IT, NOT MUCH BEING DONE!
Reed D

As much as a quarter of a flexible foam block can be wasted in downstream processing into finished products. Thanks to the efforts of process technologists and engineers, this scrap material can be recycled by at least 17 basic methods. However, only a few have found significant practical applications. Most other PU scrap ends up as uncollectable domestic waste with perhaps one key exception, materials from end-of-life vehicles.

WESTERN EUROPE-GENERAL

Accession no.709465

Item 134
Polimery Tworzywa Wielkoczasteczkowe
41, No.2, 1996, p.69-74
Polish
THERMAL METHODS OF RAW MATERIALS RECYCLING OF PLASTICS WASTES
Polaczek J; Machowska Z
Moscicki I.,Industrial Chemistry Res.Inst.

The present state of technology is reviewed (mainly from German literature of 1993-4) in the field of three principal thermal methods used for plastics wastes, namely pyrolysis (high-temperature carbonisation, coking), hydrocracking and gasification. 36 refs. Articles from this journal can be requested for translation by subscribers to the Rapra produced International Polymer Science and Technology.

EASTERN EUROPE; POLAND

Accession no.706903

Item 135
Plasticheskie Massy (USSR)
No.6, 1995, p.37-9
Russian
TECHNO-ECONOMIC ANALYSIS OF THE PYROLYSIS OF POLY(METHYL METHACRYLATE)
Solopov I V

The purpose of the study was to determine the optimum conditions of operation of pyrolysis equipment by the combined solution of equations relating to the technological and economic analysis of the process. The material considered was poly(methyl methacrylate) one of the most popular types of plastic waste. Articles from this journal can be requested for translation by subscribers to the Rapra produced International Polymer Science and Technology.

RUSSIA

Accession no.706681

Item 136
Polymer Degradation and Stability
62, No.3, 1998, p.513-21
TERTIARY RECYCLING OF POLYETHYLENE: MECHANISM OF LIQUID PRODUCTION FROM POLYETHYLENE BY THERMOLYSIS/ REACTIVE DISTILLATION
McCaffrey W C; Cooper D G; Kamal M R
McGill University

The potential of thermolysis/reactive distillation as a process for tertiary recycling of PE was investigated. The reactions were carried out at 440C and under a nitrogen atmosphere. Processing under these conditions allowed for a higher conversion of PE into a condensable liquid product than was usually reported with conventional pyrolysis. The products obtained were characterised by a high concentration of straight-chain alpha-olefins. The relative importance of the various steps in the mechanism of thermal degradation of PE was found to be very different from that observed in other studies employing only thermolysis or pyrolysis. In addition to molec.wt. reduction by random scission, intermolecular and intramolecular hydrogen transfer steps were found to be very important to the production of liquid. Intramolecular hydrogen transfer followed by beta-scission was also found to be an important mechanism for the production of short chain length alpha-olefins. 25 refs.

CANADA

Accession no.706584

Item 137
Polymer Degradation and Stability
62, No. 1, Oct. 1998, p.1-7
CHEMICAL RECYCLING OF POLYTETRAFLUOROETHYLENE BY PYROLYSIS
Simon C M; Kaminsky W
Hamburg,University

Production wastes of different PTFE compounds (containing carbon black, glass fibres and bronze) were pyrolysed in a fluidised bed reactor to gain fluorocarbons. The process parameters were varied to give high amounts of tetrafluoroethylene and hexafluoroethylene, which are important monomers for the production of fluoropolymers. At a pyrolysis temperature of 555C, yields of 76 wt% tetrafluoroethylene, 7.1 wt% hexafluoropropene and 5.2 wt% cycloperfluorobutane based on the PTFE input were obtained. At a temperature of 600C, the compounds containing carbon black and glass fibre showed a similar product distribution. For the bronze compound the yields of tetrafluoroethylene, hexafluoroethylene and cycloperfluorobutane were significantly reduced at the same temperature. The almost complete conversion of PTFE to gases by pyrolysis affords an excellent feedstock for a tetrafluoroethylene production plant and can be used to give monomers for the production of fluoropolymers.

The process is patented by Hoechst AG and a 400 t/year pilot plant is planned. 14 refs.

HOECHST AG

EUROPEAN COMMUNITY; EUROPEAN UNION; GERMANY; WESTERN EUROPE

Accession no.704604

Item 138

Kunststoffe Plast Europe

85, No.2, Feb.1995, p.16-23

FEEDSTOCK RECYCLING OF WASTE PLASTICS

Gebauer M

Plastics essentially result from raw materials which are obtained from the fossil petroleum via various stages of treatment. Owing to their adaptability and flexibility, plastics are intelligent materials with a high price/performance ratio. It is not least their efficiency which they have been widely accepted for and which has been the basic requirement for the development of many high-technology products. Notwithstanding modification of their elementary composition and many of their structural elements, however, plastics retain a high similarity to the petroleum components. Being organic compounds, plastics - in terms of their applications - therefore range among the fossil raw materials coal, petroleum, natural gas and recoverable waste plastics. Aspects described include reuse in refineries, reclamation of waste plastics, feedstock recycling, refinery methods, coal upgrading methods, special processes and feedstock recycling of plastics within DSD.

EUROPEAN COMMUNITY; EUROPEAN UNION; GERMANY; WESTERN EUROPE

Accession no.704424

Item 139

Plasticheskie Massy (USSR)

No.2, 1995, p.25-6

Russian

RECYCLING OF INJECTION-MOULDED POLYURETHANE SCRAP IN THE PRODUCTION OF SATURATED POLYESTER RESINS

Susorov I A; Kuzmitskii G E; Semenov L S

Methods are described for depolymerisation of polyurethane injection-moulding waste materials by high-temperature hydrolysis or glycolysis. One variation involves the addition of injection-moulding polyurethane scrap to the reaction mass for the synthesis of polyesters from dicarboxylic acids and glycols. During the synthesis the polyurethane is depolymerised under the action of the glycols and water formed as a result of polycondensation and can then take part in the reaction of polycondensation to form a resin of ester structure with end hydroxyl groups. It was shown that this method can be used for modification of saturated polyester resin PS-01, used as as a thermoplastic binder in the composition PTS-1 used for road-marking coatings. The effect of the content of polyurethane scrap materials on the adhesion, abrasion properties, flexibility and whiteness of composition PTS-1 is given. 8 refs. Articles from this journal can be requested for translation by subscribers to the Rapra produced International Polymer Science and Technology.

RUSSIA

Accession no.704177

Item 140

Ends Report

No.284, Sept.1998, p.20

ICI PILOT PLANT BRINGS RECYCLING OF PU FOAMS A STEP CLOSER

This article discusses a pilot plant to test the feasibility of a chemical recycling process for PU foam developed by ICI. The plant, which is in Hertfordshire, UK, will be opened in October 1998. It uses split-phase glycolysis for recycling. Full details are given.

ICI POLYURETHANES

ASIA; EUROPE-GENERAL; EUROPEAN COMMUNITY; EUROPEAN UNION; LATIN AMERICA; UK; USA; WESTERN EUROPE

Accession no.703421

Item 141

Plast' 21

No.66, Dec.1997, p.66-7

Spanish

RECYCLING OF ELECTRICAL AND ELECTRONIC EQUIPMENT

A pilot study initiated by BASF and Zentralverband der Elektrotechnishen Industrie to investigate the pyrolytic recycling of plastics and metals from electrical and electronic equipment is examined.

BASF AG; ZENTRALVERBAND DER ELEKTROTECHNISCHEN INDUSTRIE; APME; BERLIN CONSULT

BELGIUM; EUROPEAN COMMUNITY; EUROPEAN UNION; GERMANY; WESTERN EUROPE

Accession no.702553

Item 142

Journal of the National Institute of Materials and Chemical Research

6, No.4, 1998, p.159-67

Japanese

SYNTHESIS OF HIGH POLYMERS USING C1 COMPOUNDS

Masuda T; Ishigami Y; Sakaguchi H

Technologies for the sustained manufacture of high polymers that reduce the burden on the natural environment are discussed. A review is included of high polymer synthesis using syngas and its derivatives that can be obtained through steam treatment of diverse carbon resources which are available in sustainable volumes, such as waste plastics from municiple refuse. The article also introduces syngas-aided synthesis of biodegradable plastics.

JAPAN

Accession no.701590

Item 143

Revista de Plasticos Modernos

74, No.493, July 1997, p.29-42

Spanish

PVC: INCINERATION, PYROLYSIS, CHEMICAL RECYCLING, ENERGY RECOVERY, FLAMMABILITY AND TOXICITY

Barrales-Rienda J M

Instituto de Ciencia y Tecnologia de Polimeros

A survey is made of methods for the recycling of PVC and mixed waste containing PVC, including incineration with energy recovery, pyrolysis and chemical recycling. Consideration is also given to the flammability of PVC and the toxicity of its combustion products. 82 refs.

EUROPEAN COMMUNITY; EUROPEAN UNION; SPAIN; WESTERN EUROPE; WORLD

Accession no.698913

Item 144

Chemical Marketing Reporter

254, No.3, 20th July 1998, p.3/16

EVERGREEN RECYCLING IS BREAKING GROUND FOR NYLON RECLAMATION

Tullo A

Evergreen Nylon Recycling, a DSM Chemicals North America/AlliedSignal Chemical Intermediates joint venture, broke ground recently for a nylon recycling plant at DSM's site in Augusta, Ga. The plant will produce 45,000 metric tons per year of merchant grade caprolactam from used carpets. DSM may also build a carpet recycling plant in Europe, most likely in the Netherlands.

EVERGREEN NYLON RECYCLING LLC

USA

Accession no.698776

Item 145

Enercomp '95 International Conference on Composite Materials and Energy. Conference proceedings.

Montreal, Canada, 8th-10th May 1995, p.449-56. 627-63Ene

RECOVERY OF VALUABLE CHEMICAL FEEDSTOCKS FROM WASTE AUTOMOTIVE PLASTICS VIA PYROLYSIS PROCESSES

Shen Z; Day M; Cooney D

Canada,National Research Council

(Canadian Association for Composite Structures & Materials)

Each year in North America over nine million scrap vehicles are shredded to recover approximately ten million tons of ferrous metal. The process also produces three million tons of waste known as automobile shredder residue (ASR), which consists of plastics, rubber, foams, textiles, glass, dirt, rust, etc. This waste is currently landfilled. The results obtained in three different pyrolysis processes, when ASR is used as the pyrolysis feedstock, are presented. The pyrolysis processes examined include: a fast pyrolysis process, featuring rapid heat transfer and short residence times; a screw kiln unit, characterised by slow heating and long residence times; and a benchscale autoclave reactor which, in the presence of water, produces a pyrolysis liquid containing large quantities of oxygenated hydrocarbons. 7 refs.

CANADA

Accession no.698330

Item 146

Journal of Applied Polymer Science

70, No.1, 3rd Oct.1998, p.129-35

CHEMICAL RECYCLING OF FLEXIBLE PVC BY OXYGEN OXIDATION IN NAOH SOLUTIONS AT ELEVATED TEMPERATURES

Yoshioka T; Furukawa K; Sato T; Okuwaki A

Tohoku,University

The oxidative degradation of a flexible PVC pellet with oxygen was carried out in 1-25 mol/kg(m)-water sodium hydroxide solutions, at 150-260C and an oxygen partial pressure of 1-10 MPa. Dehydrochlorination of flexible PVC occurred first, followed by oxidation. The main products were oxalic acid, a mixture of benzenecarboxylic acids and carbon dioxide. One kg of flexible PVC yielded 320g of oxalic acid and 130g of benzenecarboxylic acids (as phthalic acid) under conditions of a 15m sodium hydroxide solution at 250C and an oxygen partial pressure of 5 MPa for 5 h. 19 refs.

JAPAN

Accession no.696848

Item 147

Journal of Applied Polymer Science

69, No.12, 19th Sept.1998, p.2311-9

GLYCOLYSIS OF POLYETHYLENE TEREPHTHALATE WASTES IN XYLENE

Guclu G; Kasgoz A; Ozbudak S; Ozgumus S; Orbay M

Istanbul,University; Erzacibasi San.Ve Tic.A.S.

Zinc acetate-catalysed glycolysis of waste PETP was carried out with ethylene or propylene glycols, molar ratios 1:0.5-1:3, in xylene at 170-245C. During the multiphase reaction, depolymerisation products transferred to the xylene medium from the dispersed PETP/glycol droplets, shifting the equilibrium to glycolysis. The best results were obtained from the ethylene glycol(EG) reaction at 220C, which yielded 80 mol % bis-2-hydroxyethyl terephthalate monomer and 20 mol % dimer fractions in quite pure crystalline form. Other advantages of the use of xylene in glycolysis of PETP included improvement of mixing at high PETP/EG ratios and the recycling possibility of excess glycol, which separated from the xylene phase at low temps. 15 refs.

TURKEY

Accession no.695392

Item 148
Materie Plastiche ed Elastomeri
No.6, June 1997, p.421/30
Italian
RECYCLING OF PETP BOTTLES
Meccarelli F; Roncaglia M; Maltese P

Processes for the mechanical and chemical recycling of PETP bottles and other containers are described, and waste separation techniques are also examined. 9 refs.

ASSORIMAP; PETCORE; REPLASTIC; REKO BV; PURE TECH INTERNATIONAL INC.; DOW CHEMICAL CO.

EU; EUROPEAN COMMUNITY; EUROPEAN UNION; ITALY; NETHERLANDS; USA; WESTERN EUROPE; WESTERN EUROPE-GENERAL

Accession no.695289

Item 149
ARC '97: Information to Grow the Plastics Recycling Industry. Retec proceedings.
Chicago, Il., 5th-7th Nov.1997, p.125-37. 8(13)
POST CONSUMER CARPET RECYCLING AT ALLIEDSIGNAL
Levy A B; Sifniades S
AlliedSignal Corp.
(SPE,Plastics Recycling Div.)

The recycling of nylon 6 is not new. There are two general approaches to the utilisation of nylon 6 waste streams. Chemical reprocessing usually leads to caprolactam or sometimes e-aminocaproic acid. Reprocessing based on textile technology, and other non-chemical means, leads to recycled nylon. The choice of whether to recycle to nylon 6 or caprolactam is somewhat dependent on the proposed end use. In mechanical carpet recycling there is a tremendous amount of work and therefore money required to separate the face fibre. The higher the purity required, the more expensive the process becomes. Recycling of nylon may be cost-effective, except that removal of contaminants such as dyes and coatings from polymers can be very difficult. Details are given. 2 refs.

USA

Accession no.694530

Item 150
Plastics News International
July 1998, p.18-9
WASTE TO ENERGY - A VIABLE RECYCLING OPTION
Kettle M

A feasibility study has been carried out by EPI Asia on behalf of PACOA's Plastics Environment Council into the viability of energy recovery from municipal waste. It is claimed to have demonstrated that a 250,000 tpa plant processing from a population catchment of 500,000, would generate a new 15MWe of electricity, and that given a gate fee of 50 Australian dollars per tonne for delivered waste, and power sales of 6 cents per Kwh, the plant would be viable. The recycling process is described.

EPI ASIA

AUSTRALIA

Accession no.693193

Item 151
Antec '98. Volume III. Conference proceedings.
Atlanta, Ga., 26th-30th April 1998, p.2942-5. 012
FEEDSTOCK RECYCLING OF POLYMETHYL METHACRYLATE (PMMA) BY DEPOLYMERISING IN A REACTIVE EXTRUSION PROCESS
Breyer K; Michaeli W
IKV
(SPE)

PMMA can be depolymerised by using thermal energy. A twin-screw extruder is used to heat up PMMA beyond depolymerisation temperature. The major advantages are efficient energy input through shear energy and thermal energy. The process can be operated continuously because residues are carried out of the extruder by self-wiping screws. This allows processing contaminated post consumer PMMA as well as PMMA mixed with other polymers. Furthermore, the process is accelerated by using additives so the mass throughput of the extrusion process is maximised. 7 refs.

EUROPEAN COMMUNITY; EUROPEAN UNION; GERMANY; WESTERN EUROPE

Accession no.692714

Item 152
Chemie-Ingenieur-Technik
70, No.3, March 1998, p.233-45
German
CHEMICAL RECYCLING OF PLASTICS
Sasse F; Emig G
Friedrich-Alexander University

On completion of the first life cycle of plastics, various recycling processes are available for further utilisation of these valuable materials. The choice of process will depend upon the materials to be recycled. In chemical recycling polymers are degraded to basic chemical substances which can be reused in the petrochemical industry. This route plays a key role for soiled waste plastics or waste plastics which could not previously be recycled. The pyrolysis of acrylic polymers provides a good basis for comparing a fluidised bed reactor and a tubular reactor with regard to reactor modelling. The tubular reactor with internal mass transport is a simplified model for a rotary kiln. Parameters relevant for reactor design and scale-up are presented.

EUROPEAN COMMUNITY; EUROPEAN UNION; GERMANY; WESTERN EUROPE

Accession no.692479

Item 153

Modern Plastics International

28, No.5, May 1998, p.72-4

RECYCLING OF ENGINEERING PLASTICS ADVANCES

Jones R F; Baumann M H

Franklin Polymers Inc.; GH Associates

Edited by: Kaplan W A

This article considers the recent advances in the recycling of engineering thermoplastics, in particular closed-loop partnerships, chemical recycling, and resin identification when sorting. It also examines how the automotive industry is a global target for recycling, and how business/consumer products show great recycling potential.

GENERAL MOTORS; FORD MOTOR CO.; DUPONT; AMERICAN PLASTICS COUNCIL; GE PLASTICS; APPLIANCE RECYCLING CENTERS OF AMERICA

EUROPEAN COMMUNITY; EUROPEAN UNION; GERMANY; JAPAN; USA; WESTERN EUROPE

Accession no.691800

Item 154

Revista de Plasticos Modernos

73, No.490, April 1997, p.376-7

Spanish

PETRETEC: NEW TECHNOLOGY FOR PETP RECYCLING

Details are given of the Petretec chemical recycling process developed by Du Pont for the recovery of dimethyl terephthalate and ethylene glycol from PETP waste.

DU PONT DE NEMOURS E.I.,& CO.INC.; DUPONT FILMS; VACUUM DEPOSITING INC.

USA

Accession no.691343

Item 155

Journal of Applied Polymer Science

69, No.4, 25th July 1998, p.657-65

RECYCLING OF PETP AND PVC WASTES

Lusinchi J M; Pietrasanta Y; Robin J J; Boutevin B

CEREMAP; Montpellier,Ecole Nationale Superieure de Chimie

Glycolysis of PETP leads to oligomers that are polycondensed with caprolactone. The obtained diols are extended with hexamethylene diisocyanate. In certain conditions the polyurethanes are totally miscible with PVC, leading to acceptable mechanical characteristics for the blend. A relation between the structure of the polyurethane and miscibility with PVC is described. The mechanical characteristics of the blend depends on the polyurethane chemical structure. 34 refs.

EUROPEAN COMMUNITY; EUROPEAN UNION; FRANCE; WESTERN EUROPE

Accession no.689630

Item 156

Industrial & Engineering Chemistry Research

37, No.7, July 1998, p.2582-91

POLYETHYLENE PYROLYSIS: THEORY AND EXPERIMENTS FOR MOLECULAR WEIGHT DISTRIBUTION KINETICS

Sezgi N A; Cha W S; Smith J M; McCoy B J

California,University

A novel reactor for pyrolysis of a PE melt stirred by bubbles of flowing nitrogen gas at atmospheric pressure permits uniform temperature depolymerisation. Sweep-gas experiments at temperatures 370-410 C allowed pyrolysis products to be collected separately as reactor residue (solidified PE melt), condensed vapour, and uncondensed gas products. MWDs determined by GPC indicated that random scission and repolymerisation (crosslinking) broadened the polymer-melt MWD. 19 refs.

USA

Accession no.689466

Item 157

Polymer Recycling

3, No.1, 1997/98, p.55-9

RECYCLING OF A FRACTION OF MUNICIPAL PLASTIC WASTES DEPLETED IN CHLORINE FOR A FEEDSTOCK IN A STEAMCRACKER

Joo-Sik Kim; Kaminsky W

Hamburg,University

A fraction of plastics wastes depleted in PVC was recycled at 600, 630 and 655C in a laboratory-scale fluidised bed reactor, in order to investigate the possibility of using the product oils as a feedstock in a steamcracker. At 600C, about 60 wt % of the pyrolysis product was an oil with 20 wt % of aliphatics. In the experiment at 630C, more than 62 wt % of oil was yielded. The amount of aliphatics in the oil with 21 wt % was nearly the same as that in the study at 600C. At the highest reaction temp. of 655C, the amount of aliphatics was reduced to about 12 wt %. The oils produced could be used as feedstock in a steamcracker, especially after a pretreatment, such as extraction. 8 refs.

EUROPEAN COMMUNITY; EUROPEAN UNION; GERMANY; WESTERN EUROPE

Accession no.688692

Item 158
Plastics and Rubber Weekly
No.1746, 24th July 1998, p.6
CRACKING AHEAD WITH POLSCO

The Polsco (polymer cracking in Scotland) team acquitted itself well recently in the face of challenging questions at the first interface between the Polsco project partners and local councillors and representatives from the Scottish lowlands. The project concerns the building of plastics feedstock recycling plants in the UK, the first of which is likely to be built in the vicinity of the BP pilot cracker in Grangemouth.

BP CHEMICALS LTD.; POLSCO
EUROPEAN COMMUNITY; EUROPEAN UNION; UK; WESTERN EUROPE
Accession no.688355

Item 159
Industrial & Engineering Chemistry Research
37, No.6, June 1998, p.2316-22
DEVELOPMENT OF A CONTINUOUS ROTATING CONE REACTOR PILOT PLANT FOR THE PYROLYSIS OF POLYETHENE AND POLYPROPENE
Westerhout R W J; Waanders J; Kuipers J A M; van Swaaij W P M
Twente,University

A pilot plant for the high temperature pyrolysis of polymers to recycle plastic waste to valuable products based on rotating cone reactor (RCR) technology. The RCR used in this pilot plant, the continuous RCR was an improved version of the bench-scale RCR previously used for the pyrolysis of biomass, PE and PP. 9 refs.

EUROPEAN COMMUNITY; EUROPEAN UNION; NETHERLANDS; WESTERN EUROPE
Accession no.685283

Item 160
Industrial & Engineering Chemistry Research
37, No.6, June 1998, p.2293-300
RECYCLING OF POLYETHENE AND POLYPROPENE IN A NOVEL BENCH-SCALE ROTATING CONE REACTOR BY HIGH-TEMPERATURE PYROLYSIS
Westerhout R W J; Waanders J; Kuipers J A M; van Swaaij W P M
Twente,University

The high temperature pyrolysis of PE, PP and mixtures of these polymers was studied in a novel bench-scale rotating cone reactor to identify the optimal operating conditions for this reactor. It was shown that the effect of the sand or reactor temperature on the product spectrum obtained was large compared with the effect of other parameters, e.g. residence time. 15 refs.

EUROPEAN COMMUNITY; EUROPEAN UNION; NETHERLANDS; WESTERN EUROPE
Accession no.685282

Item 161
Plast' 21
No.58, Jan./Feb.1997, p.27/33
Spanish
ENERGY RECOVERY THROUGH COMBINED INCINERATION OF MIXED PLASTICS WASTE AND SOLID MUNICIPAL WASTE

Results are presented of a study conducted in Wurzburg, Germany, in which different levels of mixed plastics waste were incinerated together with solid municipal waste. The presence of plastics led to more stable combustion without producing measurable increases in dioxins and furans. High concentrations of plastics also gave reduced carbon monoxide and sulphur dioxide emissions.

KERNFORSCHUNGSZENTRUM KARLSRUHE GMBH; DUALES SYSTEM DEUTSCHLAND; DKR; WUERZBURG,FACHHOCHSCHULE; APME
EUROPEAN COMMUNITY; EUROPEAN UNION; GERMANY; WESTERN EUROPE
Accession no.682312

Item 162
Polymer
39, No.16, 1998 p.3841-5
ALKALI-CATALYZED METHANOLYSIS OF POLYCARBONATE. A STUDY ON RECYCLING OF BISPHENOL A AND DIMETHYL CARBONATE
Hu L-C; Oku A; Yamada E
Kyoto,Institute of Technology

The alkali-catalysed methanolysis of poly(2,2-bis(4-hydroxyphenyl)propane carbonate) (PC) in a mixture of methanol (MeOH) and toluene or dioxane was studied. The treatment of PC in meOH, with a catalytic amount of sodium hydroxide, yielded only 7% bisphenol A. Using a mixed solvent of MeOH and toluene completely depolymerised PC to give 96% free bisphenol A in solid form and dimethyl carbonate in solution. The characteristics of the catalysis are discussed together with the pseudo-first rate kinetics of the depolymerisation. The reaction conditions were investigated to facilitate the recycling of PC plastics. 17 refs.

JAPAN
Accession no.682272

Item 163
Packaging Magazine
1, No.11, 4th June 1998, p.28-9
CRACKING GOOD THEORY
Dent I; Hardman S
BP Chemicals

BP Chemical's work in feedstock recycling of plastics waste is examined, and the feasibility is discussed of the construction of a commercial chemical recycling plant in the UK. The company's Grangemouth plant, which could handle 500 tons/year of plastic waste is noted, but the logistics of waste collection and supply of waste for a larger scale operation needs to be addressed, it is reported. Details are given of a 300,000 pounds sterling study by Shanks & McEwan and Valpak for the construction of a plant in Scotland.

SHANKS & MCEWAN; VALPAK

EUROPEAN COMMUNITY; EUROPEAN UNION; UK; WESTERN EUROPE

Accession no.681377

Item 164

Modern Plastics Encyclopedia

74, No.13, 1998, p.A34-5

WAVE OF RECYCLING BUMPS INTO THE SEAWALL OF ECONOMIC REALITY

Aronhalt F; Perkins R

Aronhalt Associates; American Plastics Council

Approximately 1.61 billion lb of post-consumer plastics packaging were recycled in the US in 1996, an increase of 6.4% over 1995. Similar gains are expected to be reported in Europe and Japan. In Europe, about 9.2% of total waste plastics in 1995 were recovered through mechanical or feedstock recycling methods. Waste-to-energy incineration accounted for about 16.8% of total value recovery. In the US, a consortium is sponsoring the Vehicle Recycling Development Center's research into automotive recycling technologies.

WORLD

Accession no.680755

Item 165

Kunststoffe Plast Europe

88, No.2, Feb.1998, p.32-4

PARAFFIN WAXES FROM PLASTICS WASTE

Gebauer M

Leuna-Werke

The Parak process for the recycling of polyolefins and the production of raw materials for the production of paraffin waxes, is described. The process is claimed to provide a link between mechanical and feedstock recycling, employing elements of feedstock recycling, e.g. melting and cracking. The main product obtained is paraffin wax, which can be used for coatings for cardboard and paper, and corrosion protection.

EUROPEAN COMMUNITY; EUROPEAN UNION; GERMANY; WESTERN EUROPE

Accession no.671874

Item 166

Polimeros: Ciencia e Tecnologia

6, No.3, July/Sept.1996, p.11-7

Portuguese

CHALLENGE CALLED ISO 14000

Correa C A; Pepino E; Manrich S; Zanin M; Leao A L

Sao Carlos,Universidade Federal; EMBRAPA; UNESP

The ISO 14000 series of environmental standards and their implications for the plastics industry are discussed. Aspects of ecolabelling and life cycle analysis and different options for recycling and waste disposal are examined.

STUTTGART,UNIVERSITY; EWVK; INTERNATIONAL STANDARDS ORGANISATION; ASSOCIACAO BRASILEIRA DE NORMAS TECNICAS

BRAZIL; EUROPEAN COMMUNITY; EUROPEAN UNION; GERMANY; WESTERN EUROPE; WORLD

Accession no.670825

Item 167

Revue Generale des Caoutchoucs et Plastiques

No.752, Sept.1996, p.30/4

French

INNOVATIVE ROUTE TO THE CHEMICAL RECYCLING OF PETP

The Recopet process for the chemical recycling of PETP is described. The three-stage process, which allows the recovery of highly pure terephthalic acid, commences with continuous saponification followed by chromatographic purification with activated carbon and a final acidification step.

POLYPHENIX FRANCE; ECO-EMBALLAGES SA

EUROPEAN COMMUNITY; EUROPEAN UNION; FRANCE; WESTERN EUROPE

Accession no.670793

Item 168

Polymer Recycling

2, No.4, 1996, p.309-15

THERMAL AND CATALYTIC DEGRADATION OF MUNICIPAL WASTE PLASTICS INTO FUEL OIL

Sakata Y; Uddin A; Muto A; Koizumi K; Narazaki M; Murata K; Kaji M

Okayama,University; Mitsui Engineering & Shipbuilding Co.Ltd.; Plastic Waste Management Institute

This comprehensive article supplies details of a new catalytic process for the degradation of municipal waste plastics in a glass reactor. The degradation of plastics was carried out at atmospheric pressure and 410 degrees C in batch and continuous feed operation. The waste plastics and simulated mixed plastics are composed of polyethylene, polypropylene, polystyrene, polyvinyl chloride, acrylonitrile butadiene styrene, and polyethylene terephthalate. In the study, the degradation rate and yield of fuel oil recovery promoted by the use of silica alumina catalysts are compared with the non-catalytic thermal degradation. 9 refs.

JAPAN

Accession no.668719

Item 169
Polymer Recycling
2, No.4, 1996, p.291-97
MIXED COMBUSTION OF AUTOMOTIVE SHREDDER RESIDUES WITH MUNICIPAL SOLID WASTE: A SOUND ROUTE TO ENERGY RECOVERY FROM END OF LIFE VEHICLES
Jean A A
Elf Atochem

This comprehensive article describes the advantages of using plastic in automotive applications. Reducing weight in vehicles makes them more economical and environmentally friendly by reducing fuel consumption and gas emissions. The article focuses on a number of trials of three recovery options for plastics from end of life vehicles: mechanical recycling, feedstock or chemical recycling and clean waste-to-energy incineration. The article supplies details of trials of possible processes for energy recovery from shredding refuse by means of co-combustion, one of the most promising being the co-combustion of shredding residues together with municipal solid waste.

EUROPEAN COMMUNITY; EUROPEAN UNION; FRANCE; WESTERN EUROPE

Accession no.668716

Item 170
Plastics and Rubber Weekly
No.1722, 6th Feb.1998, p.9
INDUSTRY PARTNERSHIPS GIVE NEW LEASE OF LIFE TO LCA

ICI Acrylics believes that greater cooperation between companies and a revised approach to life cycle analysis are the keys to the industry's future environmental sustainability. The company has invested over 2m pounds sterling in an on-going monomer recovery project, which encompasses a joint research programme with Mitsubishi Rayon. The project focuses on increasing the efficiency of acrylic depolymerisation and overcoming technical issues such as its use in recycling flame retardant acrylics.

ICI ACRYLICS
EUROPEAN COMMUNITY; EUROPEAN UNION; UK; WESTERN EUROPE

Accession no.668049

Item 171
Ends Report
No.275, Dec.1997, p.27-8
ICI SEES MARKET ADVANTAGE THROUGH ACRYLICS RECYCLING

ICI Acrylics' activities in chemical recycling of acrylics is discussed. The company is offering a take-back service for scrap PMMA which it chemically recycles back into MMA. Together with Mitsubishi Rayon, it has established a joint venture to develop more efficient depolymerisation technology which will produce a purer material enabling a higher proportion of recylate to be used in clear PMMA without affecting transparency. According to ICI, depolymerisation requires less energy than making virgin monomer from oil, and is currently conducting a life cycle analysis to calculate the level of dematerialisation which can be achieved by using different proportions of recyclates in products.

ICI ACRYLICS; DUSCHOLUX; MITSUBISHI RAYON CO.LTD.
EUROPEAN COMMUNITY; EUROPEAN UNION; JAPAN; UK; WESTERN EUROPE

Accession no.667011

Item 172
Ends Report
No.275, Dec.1997, p.14-5
PLASTIC BOTTLE RECYCLING RISES, BUT BIG GAP TO BRIDGE TO EC TARGET

The recovery rate of plastic bottles in the UK is discussed with reference to the need to meet packaging recycling targets for 2001 under the EC Directive on packaging waste. It is argued that there remains a need for the construction of several feedstock recycling plants in order to achieve the minimum 15% recycling rate for each major packaging material.

RECOUP; VALUPLAST
EUROPEAN COMMUNITY; EUROPEAN UNION; UK; WESTERN EUROPE

Accession no.667010

Item 173
Ends Report
No.275, Dec.1997, p.13-4
ENERGY RECOVERY OF PACKAGING WASTE FACES LEGAL OBSTACLE

Confusion as to what constitutes municipal waste is presenting an obstacle to the use of packaging waste as a fuel in cement kilns. Whilst cement kilns can burn hazardous waste, they cannot burn a wide range of non-hazardous materials, it is reported. The case of Castle Cement is described which planned to burn a range of non-hazardous commercial and industrial wastes. Some waste-fired combustion processes, however, such as UK Waste's Fibre Fuel operation have been granted derogations where fuel is manufactured by advanced mechanical processes, which includes the production of fuel pellets. This latter process would be pointless for the cement industry since their fuels have to be pulverised. The problems are further discussed with reference to current European legislation.

UK,ENVIRONMENT AGENCY; CASTLE CEMENT; UK WASTE
EUROPEAN COMMUNITY; EUROPEAN UNION; UK; WESTERN EUROPE

Accession no.667009

Item 174
Journal of Applied Polymer Science
67, No.2, 10th Jan.1998, p.341-8
EFFECTS OF CATALYST ACTIVITY AND HZSM-5 CHANNEL VOLUME ON POLYPROPYLENE CRACKING
Negelein D L; Rong Lin; White R L
Oklahoma,University

Effects of catalyst acidity and the restricted reaction volume afforded by HZSM-5 on the catalytic cracking of PP in waste recycling are described. PP cracking by silica-alumina and HZSM-5 catalysts yielded olefins as primary volatile products. In addition, HZSM-5 channels restricted carbenium ion rearrangements and facilitated formation of significant amounts of propylene and alkyl aromatic volatile products. The higher acidity of sulphated zirconia compared with the other catalysts resulted in an increase in the frequency of hydride abstractions, resulting in the formation of significant yields of saturated hydrocarbons and organic residue for this catalyst. Primary PP cracking products could be derived from carbenium ion reaction mechanisms. 20 refs.

USA

Accession no.665757

Item 175
Chemical Marketing Reporter
252, No.26, 29th Dec.1997, p.1/8
DEPOLYMERISATION GETS THE NOD AS ROUTE TO LOW-COST FEEDSTOCK
Brand T

Large companies are taking a closer look at depolymerising nylon, polyester and PU products that would otherwise end up in landfills. The most recent project is a worldscale facility to be built by DSM Chemicals North America and AlliedSignal that will produce about 100 million pounds of caprolactam per year by depolymerising nylon 6. The facility will remove about 200 million pounds of carpet from landfills annually. DuPont says it has a process, ammonolysis, that can depolymerise a combination of nylon 6 and 66.

USA

Accession no.664908

Item 176
Informations Chimie
No.378, May 1996, p.107-9
French
RECOPET PROCESS: AN INNOVATION IN THE CHEMICAL RECYCLING OF PETP

An account is given of the Recopet process for the chemical recycling of PETP and its use in a pilot plant operated by Tredi and Polyphenix France at Tessenderlo in Belgium. The prepared waste is subjected to a continuous saponification process, followed by chromatographic purification with activated carbon and a final acidification stage, resulting in the recovery of highly pure terephthalic acid.

EMC; POLYPHENIX FRANCE; TREDI; TESSENDERLO CHEMIE SA; INSTITUT FRANCAIS DU PETROLE

BELGIUM; EUROPEAN COMMUNITY; EUROPEAN UNION; FRANCE; WESTERN EUROPE

Accession no.663540

Item 177
Journal of the National Institute of Materials and Chemical Research
5, No.4, 1997, p.175-183
Japanese
APPLICATION OF SUPERCRITICAL FLUIDS TO EARTH ENVIRONMENT PROTECTION TECHNOLOGY
Sako T; Sugeta T

Three kinds of promising innovative environmental applications using supercritical fluids to solve problems of energy, resources and global environment are shown: the complete decomposition of hazardous compounds with supercritical water, the recycling of waste plastics with supercritical methanol or supercritical water, and chemical reaction in supercritical carbon dioxide which is free from toxic organic solvents. Supercritical methanol depolymerised waste condensation polymers such as PETP and PEN into constituent monomers easily. The supercritical water decomposed composite plastics such as FRP into the glass fibre and fuel oil with high efficiency. 12 refs.

JAPAN

Accession no.662847

Item 178
Industrial & Engineering Chemistry Research
36, No.11, Nov.1997, p.4523-9
TRANSFORMATION OF SEVERAL PLASTIC WASTES INTO FUELS BY CATALYTIC CRACKING
Arandes J M; Abajo I; Lopez-Vakerio D; Fernandez I; Azkoiti M J; Olazar M; Bilbao J
Bilbao,Universidad del Pais Vasco

The thermal and catalytic cracking of PP, PS, and SBR waste, dissolved in light cycle oil, was studied in a riser simulator. 19 refs.

EUROPEAN COMMUNITY; EUROPEAN UNION; SPAIN; WESTERN EUROPE

Accession no.660825

Item 179
Industrial & Engineering Chemistry Research
36, No.11, Nov.1997, p.4436-44
ULTRAPYROLYTIC UPGRADING OF PLASTIC WASTES AND PLASTICS/HEAVY OIL MIXTURES TO VALUABLE LIGHT GAS PRODUCTS
Lovett S; Berruti F; Behie L A
Calgary,University

Viable operating conditions were identified experimentally for maximising the production of ethylene, propylene, styrene and benzene from the pyrolysis of waste products. Data are given for pyrolysis temperature, product reaction time, and quench time using a batch microreactor and a pilot-plant-sized reactor. 26 refs.

CANADA

Accession no.660824

Item 180
Kunststoffe Plast Europe
87, No.11, Nov. 1997, p.58-60
RECYCLING TECHNOLOGIES
Schalles H
IKR

Development work in plastics recycling is concentrated on the processing of clean, high quality wastes with specialised, tailor-made equipment, with a clear trend towards a marketable quality product and using quality assurance systems. Particular details are given of PETP recycling, size reduction lines, agglomeration and cleaning, separation and sorting, plastication in the extruder, melt filtration, recycling of commingled plastics and feedstock recycling.

EUROPEAN COMMUNITY; EUROPEAN UNION; GERMANY; WESTERN EUROPE

Accession no.660543

Item 181
Plast' 21
No.48, Dec.1995, p.47-9
Spanish
CHEMICAL RECYCLING, ANOTHER COMPONENT IN AN INTEGRATED SYSTEM

Chemical recycling is examined as a means for plastics waste management, and commercial developments by a number of West European companies are described.

VEBA OEL; RWE; RHEINBRAUN AG; BASF AG; DSM NV; BP CHEMICALS LTD.; ENICHEM; ELF ATOCHEM SA; STAHLWERKE BREMEN; PETROFINA SA

BELGIUM; EU; EUROPEAN COMMUNITY; EUROPEAN UNION; FRANCE; GERMANY; ITALY; NETHERLANDS; UK; WESTERN EUROPE; WESTERN EUROPE-GENERAL

Accession no.659588

Item 182
International Journal of Polymeric Materials
37, Nos.3-4, 1997, p.173-99
CHEMICALS AND ENERGY FROM MEDICAL POLYMER WASTES. II. MALEATED PYROLYSIS PRODUCTS IN IPP/LLDPE PROCESSING
Vasile C; Deanin R D; Mihaies M; Roy C; Chaala A; Ma W
Massachusetts,University; CEPROPLAST SA; Quebec,Universite Laval

Details are given of ways of obtaining energy from the pyrolysis of disposable syringes. The waxy product was chemically modified with maleic anhydride and tested in the processing of PP/LLDPE blends in a rheometer or twin-screw extruder. Compatibility of components in binary or ternary blends and DSC results are discussed. 35 refs.

CANADA; EASTERN EUROPE; RUMANIA; USA

Accession no.656211

Item 183
Revista de Plasticos Modernos
71, No.477, March 1996, p.290/301
Spanish
CHEMICAL RECYCLING OF PLASTICS
Vargas L
Repsol SA

The origins and composition of plastics wastes and factors affecting their recycling are discussed. Particular attention is paid to chemical recycling and incineration, with and without energy recovery, and a number of developments in chemical recycling techniques are examined. 19 refs.

EUROPEAN COMMUNITY; EUROPEAN UNION; SPAIN; WESTERN EUROPE; WESTERN EUROPE-GENERAL

Accession no.649948

Item 184
Designed for life:Composites '94. Publication number 293/9. Conference proceedings.
London, BPF, 1994, p.59-64. 627
STRATEGIES FOR RECYCLING AND ENERGY RECOVERY FROM THERMOSET COMPOSITES
Pickering S J; Bevis M J; Hornsby P R
Nottingham,University; Brunel University
(BPF)

There are a number of routes for recovering and recycling of materials from thermoset composites. With the exception of the ERCOM project none of these are particularly well developed. The most promising for development are innovative comminution techniques for the preparation of relatively uncontaminated forms of scrap as a reinforcement for development of high value products based on thermoset, thermoplastic and elastomeric matrices and energy recovery processes combined with recovery and recycling of incombustible materials, particularly fibres, dealing with contaminated and mixed forms of scrap. This paper considers the potential of these approaches by looking at previous work done in this area and a new collaborative project between the University of Nottingham, Brunel University and sixteen industrial companies funded by the LINK Structural Composites Programme. The comminution task will be led by Brunel University and the combustion task by the University of Nottingham. 12 refs.

EUROPEAN COMMUNITY; EUROPEAN UNION; UK; WESTERN EUROPE

Accession no.649748

Item 185
European Plastics News
24, No. 8, Sept. 1997, p.37
PERPETUAL MOTION
Lee M

Details are given of DuPont's Petretec process for the regeneration of PETP. Although PETP is currently one of the most widely recycled plastics, existing technologies are unable to deal with impurities, it is stated. The Petretec process is able to handle polyester with a variety of contaminants at levels up to 8-10%. The process chemically regenerates scrap PETP, taking it back to its constituent dimethyl terephthalate and ethylene glycol molecules, and enables it to be used in first grade high quality applications.

DUPONT
USA

Accession no.649274

Item 186
Patent Number: US 5556890 A 19960917
RECLAIMING EPSILON-CAPROLACTAM FROM CARPET WASTE
Halderit A H T; Booij M; Hendrix J A J; Frentzen Y H
DSM NV

The present invention is for a method of preparing purified epsilon-caprolactam from carpet waste containing nylon-6. Contaminated epsilon-caprolactam is obtained by depolymerising nylon-6 from carpet waste. A mixture of the contaminated epsilon-caprolactam and water is then hydrogenated in the presence of hydrogen and a hydrogenation catalyst in order to produce purified epsilon-caprolactam. The amount of contaminated epsilon-caprolactam in the mixture can be between 10 and 95% by weight.

EUROPEAN COMMUNITY; EUROPEAN UNION; NETHERLANDS; WESTERN EUROPE

Accession no.645161

Item 187
Warmer Bulletin
No.54, May 1997, p.16-7
ROLE OF PLASTICS IN ENERGY RECOVERY
Thurgood M

This article discusses the role of plastics in energy recovery, and highlights several full scale investigations that have taken place in the UK, the USA, Japan, Sweden, Finland, and Germany. A summary of the results of the studies, and a conclusion, are provided.

EBARA CORP.; UMEA,UNIVERSITY
CANADA; EUROPE-GENERAL; EUROPEAN COMMUNITY; EUROPEAN UNION; FINLAND; GERMANY; JAPAN; SCANDINAVIA; SWEDEN; USA; WESTERN EUROPE

Accession no.639530

Item 188
Industrial & Engineering Chemistry Research
36, No.4, April 1997, p.1373-83
CHEMICAL RECYCLING OF POLY(ETHYLENE TEREPHTHALATE)
Paszun D; Spychaj T
Szczecin,Technical University

This paper reviews the state of the art in the field of chemical recycling of PETP. Advantages of the chemical recycling of PETP, the theoretical basis of the ester bond cleavage, and a wide spectrum of degrading agents and final products are presented. Chemical processes applied in polymer recycling are divided into six groups, methanolysis, glycolysis, hydrolysis, ammonolysis, aminolysis, and other methods. Numerous possibilities for the utilisation of waste PETP as a very useful raw chemical material are described on the basis of literature. Examples include coatings, plasticisers and low grade PU for use in putties and sealants. A comparison of chemical recycling methods is carried out. The following aspects were taken into consideration, (i) flexibility in utilising a variety of waste types, (ii) conditions necessary for degradation including safety requirements, (iii) economic aspects, and (iv) product versatility. Citations include 46 patents. 108 refs.

EASTERN EUROPE; POLAND

Accession no.639169

Item 189
European Chemical News (Chemscope)
May 1997, p.12
LINDE, A CLASS OF ITS OWN
Williams D

The success of Linde's process and engineering contracting business is discussed. Its proprietary technology includes petrochemical projects, largely ethylene crackers, which account for roughly half of its sales, and over 300 processes protected by 1500 patents.

Air separation, syngas and the rapidly growing hydrogen segment, pharmaceutical and environmental projects are included in its portfolio.

LINDE AG

WORLD

Accession no.634892

Item 190

Polymer

38, No.9, 1997, p.2281-5

RECYCLING OF UNSATURATED POLYESTER RESIN USING PROPYLENE GLYCOL

Yoon K H; DiBenedetto A T; Huang S J
Connecticut,University

Cured unsaturated polyester resin was degraded using propylene glycol at different temperatures and the resulting material was reacted with maleic anhydride to make recycled resin. The curing reaction for the recycled resin was faster than that for the neat resin. The mechanical properties of mixtures of neat and recycled resins were measured and the mixture containing 90 wt% neat and 10 wt% recycled resin showed the highest impact strength. Adding 15 wt% of chopped glass fibre to the recycled resin increased the impact strength further. 13 refs.

USA

Accession no.634475

Item 191

New Scientist

153, No.2072, 8th March1997, p.6

RECYCLING MADNESS REVIVES OBSOLETE PLANTS

Charles D

This article describes the re-opening of several old chemical plants, previously used to convert coal into oil or gas, but now being resurrected to gasify recycled plastics. German recycling law requires businesses to collect and recycle most commercial packaging, and the chemical recycling of plastics attracts substantial financial subsidies, although environmentalists argue that little money is being spent on preventing the creation of waste.

EUROPEAN COMMUNITY; EUROPEAN UNION; GERMANY; WESTERN EUROPE

Accession no.632553

Item 192

Ludwigshafen, 1996, pp.6. 12ins. 7/3/97. Technical Information for Experts, No.08/96e

PLASTICS RECYCLING - ACTIVITIES OF BASF

BASF AG

The plastic recycling activities of BASF are reported and discussed. Recycling of plastics products is investigated within the company's pilot plants in cooperation with customers, and in projects which are supported by industry partners. BASF offers recycled grades for many of its engineering plastics, and also for Styropor, its expanded polystyrene bead. The company is involved in mechanical recycling, feedstock/chemical recycling, and energy recovery. Details are given of the techniques used, current projects, and a history of BASF's activities in this field is included.

EUROPEAN COMMUNITY; EUROPEAN UNION; GERMANY; WESTERN EUROPE

Accession no.629654

Item 193

Kunststoffe Plast Europe

87, No.2, Feb.1997, p.19-20

FEEDSTOCK RECYCLING OF POLYMETHYL METHACRYLATE (PMMA): DEGRADATIVE EXTRUSION IN A TWIN-SCREW EXTRUDER

Michaeli W; Breyer K
RWTH

Post-consumer PMMA plastics can be depolymerised back into their starting components. Degradative extrusion in twin-screw extruders can be used for this process. 10 refs. Translation of Kunststoffe, 87, No.2, Feb.1997, p.183-8

EUROPEAN COMMUNITY; EUROPEAN UNION; GERMANY; WESTERN EUROPE

Accession no.629123

Item 194

SAMPE Journal

32, No.5, Sept./Oct.1996, p.52-7

ADVANCED COMPOSITES RECYCLING

Unser J F; Staley T; Larsen D
Environmental Technical Services; Missouri,University

A significant amount of waste composites is generated each year and the need for a recycling method is becoming a necessity. Environmental Technical Services has developed, with the support of the University of Missouri-St.Louis, a method for recovering valuable constituents from composite materials. The process converts the polymer matrix to lower chain hydrocarbons and fuel gas leaving behind fibres. Mechanical tests of BMC panels, reinforced concrete and compression moulded panels made with recovered fibres were carried out. 10 refs.

USA

Accession no.628960

Item 195

SAMPE Journal

32, No.5, Sept./Oct.1996, p.46-51

RECYCLING PROCESS FOR SCRAP COMPOSITES AND PREPREGS

Allred R E
Adherent Technologies

Currently, scrap fibre-reinforced composite materials and prepregs are landfilled, although these materials represent a valuable resource. A novel tertiary recycling process is being developed for polymer matrix composites and prepregs where, under the action of low heat and catalysts, the matrix is converted to a mixture of low molecular weight hydrocarbons and removed from the fibres as a gas. The fibres may then be reused as reinforcements for new composites and the hydrocarbons refined and used as chemicals or fuel. 17 refs.

USA

Accession no.628959

Item 196

Advances in Automotive Plastic Components and Technology. Conference proceedings.
Detroit, Mi., 27th Feb.-2nd March 1995, p.179-85.
63Tr.Ro

RECYCLING OF THERMOSET POLYMERS

Kresta J E; Xiao H X; Cejpek I; Kytner J
Detroit,Mercy University
(Society of Automotive Engineers)

The recycling of polyurea-urethane (PU-U) reaction injection moulding (RIM) materials, which are widely used in the automotive industry, is investigated. The recycling process is based on the catalysed chemical decrosslinking, using short chain diol as a reactant. The transesterification (decrosslinking reaction) is studied in the Haake mixer and twin-screw extruder. The effect of various parameters (temperature, reactant/PU-U RIM ratio, rpm etc.), on the yield of liquid oligomers is determined. The resulting liquid oligomers are characterised using various analytical methods and used in the preparation of new products such as coatings, adhesives etc. The results show that the catalysed chemical decrosslinking of PU-U RIM can be a potential recycling method. 20 refs.

USA

Accession no.628159

Item 197

Journal of Applied Polymer Science
63, No.10, 7th March 1997, p.1287-98

ACID CATALYSED CRACKING OF POLYSTYRENE

Lin R; White R L
Oklahoma,University

Catalytic cracking of high and low molecular weight PS was carried out using silica/alumina, sulphated zirconia and zeolite catalysts. The effects of catalyst acidity and the restricted reaction volume afforded by the zeolite on the volatile cracking products were examined. Styrene was a minor cracking product. The most abundant volatile product generated was benzene, and alkyl benzenes and indanes were also detected in significant yields. Various thermal analysis techniques were used to obtain volatilisation activation energies for polymer-catalyst samples and to elucidate probable reaction pathways. The detected products were explained by reaction mechanisms which began with protonation of PS aromatic rings. 26 refs.

USA

Accession no.624418

Item 198

International Polymer Science and Technology
23, No.9, 1996, p.T/106-10

TECHNICAL AND ECONOMIC ANALYSIS OF THE PYROLYSIS OF POLYMETHYL METHACRYLATE

Solopov I V

Optimum operating conditions of the pyrolysis unit by joint solution of equations of technological and economic analysis of the process. PMMA, one of the most popular types of plastics waste, was chosen as the example. Stages of technological analysis of industrial chemical processes are presented. 7 refs. Translation of Plast.Massy, No.6, 1995, p.37

RUSSIA

Accession no.619879

Item 199

Financial Times
No.33200, 27th Jan.1997, p.8

BUSINESSES NOT PREPARED FOR RECYCLING LAW

Boulton L; Burt T

According to a study carried out by Valpak, nearly 70% of senior managers in Britain are not aware of an imminent law requiring industry to pay for the recovery or recycling of half the packaging waste it puts into circulation. Once the legislation is passed by parliament, companies will have at most six months to register with a scheme to recycle a proportion of packaging on their behalf unless they plan to do it themselves. The deadline for the recovery of 50% of packaging waste, with at least 25% recycled, is 2001.

VALPAK

EUROPEAN COMMUNITY; EUROPEAN UNION; UK; WESTERN EUROPE

Accession no.619711

Item 200

Recycle '95. Conference proceedings.
Davos, 15th-19th May 1995, paper 31. 8(13)

MIXED COMBUSTION OF AUTOMOTIVE SHREDDER RESIDUES WITH MUNICIPAL SOLID WASTE

Jean A A; Gloriod P
Elf Atochem SA
(Maack Business Services)

This paper considers: plastics and their positive role during the lifetime of a car, the case of end-of-life vehicles, energy recovery from shredding refuse by means of co-

combustion, "Cyclergie" technology, and characteristics of the Pontivy recycling plant in France.

EUROPEAN COMMUNITY; EUROPEAN UNION; FRANCE; WESTERN EUROPE

Accession no.617410

Item 201

Recycle '95. Conference proceedings.
Davos, 15th-19th May 1995, paper 20. 8(13)

FEEDSTOCK RECYCLING OF POST CONSUMER WASTE PLASTICS

Niemann K
Kohleoel-Anlage Bottrop GmbH
(Maack Business Services)

This paper provides a detailed examination of feedstock recycling of post-consumer waste plastics. The basics of liquid phase hydrogenation are explained, the so-called VCC-Technology process is described, and the recycling activities of the German company, Kohleoel-Anlage Bottrop GmbH, are highlighted.

EUROPEAN COMMUNITY; EUROPEAN UNION; GERMANY; WESTERN EUROPE

Accession no.617399

Item 202

Recycle '95. Conference proceedings.
Davos, 15th-19th May 1995, paper 19. 8(13)

POLYAMIDE 66 AND 6 CHEMICAL RECYCLING

Smith R A; Gracon B E
DuPont
(Maack Business Services)

This paper describes DuPont's very successful progress in the chemical recycling of nylon carpets by a patented ammonolysis process. Each stage of the process is described, from the collection of old and dirty used carpets, right through to the production of high-purity monomers.

USA

Accession no.617398

Item 203

Recycle '95. Conference proceedings.
Davos, 15th-19th May 1995, paper 18. 8(13)

CHEMICAL RECYCLING DILEMMA

Caruthers W C
Amoco Corp.
(Maack Business Services)

This paper focuses on the dilemmas surrounding plastics solid waste management in the USA today, providing information on the present situation and how it was reached, recycling realities, good and bad recycling, advanced recycling, and the challenge that industry now faces.

AUSTRALIA; CANADA; EUROPE-GENERAL; JAPAN; USA

Accession no.617397

Item 204

Recycle '95. Conference proceedings.
Davos, 15th-19th May 1995, paper 61. 8(13)

WASTE PLASTIC LIQUEFACTION USING THERMAL AND CATALYTIC CRACKING PROCESS

Funahashi E
Nippon Steel Corp.
(Maack Business Services)

This paper outlines waste plastic liquefaction technology, and discusses its use in the operation of a demonstration plant. The following aspects are considered: plastics suitable for liquefaction, basic principals of liquefaction technology, the liquefaction process, examples, results, and conclusions. 4 refs.

JAPAN

Accession no.617365

Item 205

Recycle '95. Conference proceedings.
Davos, 15th-19th May 1995, paper 60. 8(13)

COMBUSTION AND ENERGY RECOVERY OF WASTE PLASTIC BY INTERNALLY CIRCULATING FLUIDISED BED BOILER

Tsukamoto K; Kurihara K
EBARA Corp.
(Maack Business Services)

This paper introduces the Twin-Interchanging Fluidised Bed Incinerator (TIF) from EBARA Corp. of Japan, and describes a combustion test carried out by the company in collaboration with the Plastic Waste Management Institute, on waste plastic separated from municipal refuse, verifying the level of non-polluting combustion and high-efficiency energy recovery. The results of the test are presented, with considerations and conclusions.

JAPAN

Accession no.617364

Item 206

R'95 - Recovery, Recycling, Re-Integration. Volume I: Concepts, Life Cycle Analysis, Legal and Economic Instruments. Conference proceedings.
Geneva, 1st-3rd Sept. 1995, p.I.175-80. 8(13)

ENVIRONMENTAL IMPACTS BY DISPOSAL PROCESSES

Molgaard C; Alting L
Denmark,Technical University
Edited by: Barrage A; Edelmann X
(EMPA; Swiss Federal Laboratories for Mat.Testing & Res.)

A steadily increasing demand for the recycling of polymers has resulted in a demand for methods making it possible to compare the influence of different disposal processes on the environment and on the resources. Ranking of different disposal processes in an environmentally and resource-compatible way can be carried out by ecoprofiles. An ecoprofile is an assessment of the environmental and resource impacts for a given disposal process and those processes which are influenced by the disposal process. The use of ecoprofiles is illustrated by comparison of four different disposal scenarios: material reprocessing, pyrolysis, incineration and landfill of LDPE film. 11 refs.

DENMARK; EUROPEAN COMMUNITY; EUROPEAN UNION; SCANDINAVIA; WESTERN EUROPE

Accession no.615245

Item 207

R'95 - Recovery, Recycling, Re-Integration. Volume V: Thermal Processes, Energy Recovery. Conference proceedings.

Geneva, 1st-3rd Sept. 1995, p.V.73-80. 8(13)

ENERGY RECOVERY FROM USED PLASTICS BY GASIFICATION

De Stefanis P; Di Palo D; Velcich G; Zagaroli M

ENEA; Centro Ricerche Casaccia; Daneco Danieli Ecologia SpA

Edited by: Barrage A; Edelmann X

(EMPA; Swiss Federal Laboratories for Mat.Testing & Res.)

The results of some test runs conducted on an experimental gasification plant having a 2 MWt capacity are described. The tests were carried out with PE and PETP, from the separate collection of bottles and containers for liquids. The tests were directed at obtaining information on the applicability (both as far technical aspects and the impact on environment are concerned) of the selected gasification and related produced gas treatment plant, for the recycling of used plastics through energy recovery, according to Italian legislation in force. To this end, specific test objectives were to develop a set of material and energy balances and to detect the characteristics both of produced gas and the flue gas and other residues coming out of the plant. The test runs were performed from February to April 1993 under the supervision of ENEA (Italian National Agency for New Technology, Energy and the Environment). During test runs ENEA observed the plant operation, collected data and conducted extensive sampling of the produced gas, flue gas from endothermic engine, bottom/fly ashes from gasification and treatment section. After a short account of the plant, the activities performed, the results obtained, the problems encountered and possible future developments are described.

EUROPEAN COMMUNITY; EUROPEAN UNION; ITALY; WESTERN EUROPE

Accession no.615229

Item 208

R'95 - Recovery, Recycling, Re-Integration. Volume IV: Chemical Processes, Biological Processes, Hospital Waste. Conference proceedings.

Geneva, 1st-3rd Sept. 1995, p.IV.120-5. 8(13)

CHEMICAL DISSOCIATION OF AMIDE BONDS - RECYCLING OF PURE POLYMER WASTES

Seyfarth H E; Riedel B; Meusel E; Muller W; Taeger E

Thueringisches Institut fuer Textil- & Kunststoff-Forschung eV

Edited by: Barrage A; Edelmann X

(EMPA; Swiss Federal Laboratories for Mat.Testing & Res.)

Polymers with hetero-atoms in the chain are suitable for chemical recycling of waste materials. In addition to depolymerisation (nylon 6) and solvolysis (nylon 6,6, PETP, PU) the degradation of aliphatic polyamides with dicarboxylic acids, diamines and cyclic anhydrides, especially trimellitic anhydride, becomes more and more important. The utilisation of the obtained fragments is described.

EUROPEAN COMMUNITY; EUROPEAN UNION; GERMANY; WESTERN EUROPE

Accession no.615225

Item 209

R'95 - Recovery, Recycling, Re-Integration. Volume IV: Chemical Processes, Biological Processes, Hospital Waste. Conference proceedings.

Geneva, 1st-3rd Sept. 1995, p.IV.51-8. 8(13)

RECOVERY OF HYDROCHLORIC ACID FROM THERMAL PROCESSING OF WASTE PVC

Schaub M

Sulzer Chemtech AG

Edited by: Barrage A; Edelmann X

(EMPA; Swiss Federal Laboratories for Mat.Testing & Res.)

The recycling of plastic materials becomes more and more important, but unfortunately it is not possible to make materials from the same quality as that of the virgin material. For PVC, this problem is even more pronounced than for other plastics. A process is studied which is able to destroy the waste PVC, but which can recover the most important component of it - chlorine - as a raw product for VCM manufacture, with a very high yield. Most of the energy contained in the PVC can be recovered as electrical power and steam. 4 refs.

SWITZERLAND; WESTERN EUROPE

Accession no.615223

Item 210

Brussels, 1995, pp.2. 12 ins. 19/1/96.

PLASTICS MAKE IT POSSIBLE... TO TAKE A FLEXIBLE APPROACH TO RECOVERY WASTE, MAXIMISING ENVIRONMENTAL AND ECONOMIC GAIN

APME

Post-consumer plastic waste recycling is discussed with special reference to feedstock recycling, the advantages it has over mechanical recycling, and the techniques involved. Chemolysis and thermolysis are explained, and details are briefly given of examples of current European feedstock recycling operations.

BELGIUM; EUROPEAN COMMUNITY; EUROPEAN UNION; WESTERN EUROPE

Accession no.614309

Item 211

Brussels, 1995, pp.2. 12 ins. 19/1/96.

PLASTICS MAKE IT POSSIBLE... TO TURN WASTE INTO LIGHT AND HEAT FOR OUR HOMES, BUSINESSES AND COMMUNITY FACILITIES

APME

The recycling of plastics waste is considered with respect to energy recovery through incineration. It is claimed that burning solid municipal waste could produce nearly 10% of Europe's domestic electricity and heat and conserve resources by replacing, for example, over half of current coal imports to Western Europe. The potential for power from waste plastics and examples of energy from waste in action are described.

BELGIUM; EUROPEAN COMMUNITY; EUROPEAN UNION; WESTERN EUROPE

Accession no.614306

Item 212

Materie Plastiche ed Elastomeri

No.4, April 1995, p.186-91

Italian

CHEMICAL RECYCLING: BACK TO THE ORIGINS

Modini G

Methods used in the recycling of plastics are described, and details are given of a chemical recycling technique used by Veba Oel in its plant in Bottrop, Germany. This liquid phase hydrogenation process produces high quality synthetic oils, suitable for use as refinery feedstocks, from mixtures of vacuum distillation residues, scrap plastics and other industrial wastes.

VEBA OEL AG; APME

EUROPEAN COMMUNITY; EUROPEAN UNION; GERMANY; WESTERN EUROPE; WESTERN EUROPE-GENERAL

Accession no.611858

Item 213

R'95 - Recovery, Recycling, Re-Integration. Volume II: Design for Avoidance, Design for Recycling. Conference proceedings.

Geneva, 1st-3rd Sept. 1995, p.II.24-9. 8(13)

RECYCLING OF ENGINEERING PLASTICS - OPTIONS AND LIMITATIONS

Riess R

Bayer AG

Edited by: Barrage A; Edelmann X

(EMPA; Swiss Federal Laboratories for Mat.Testing & Res.)

Intended legislation on the recycling of end of life vehicles and electronic waste is leading to increasing recovery activities within the industries concerned. For years Bayer, as a producer of engineering plastics and PUs, has been heavily involved in numerous recycling projects focused on automotive exterior and interior parts, as well as electrical and electronic goods. Promising results refer to detection technologies, preparation and upgrading of used plastic materials and their reintroduction Into new applications. Waste management problems can only be solved if all three recycling routes - mechanical, feedstock and energy recovery - are pursued without preference or discrimination. The future has to be determined by ecological and economic considerations.

EUROPEAN COMMUNITY; EUROPEAN UNION; GERMANY; WESTERN EUROPE

Accession no.611282

Item 214

Advanced Materials Newsletter

18, No.20, Iss.411, 21st Oct.1996, p.2

TITAN TECHNOLOGIES, ADHERENT TECHNOLOGIES RESEARCH COMPOSITES RECYCLING VIA SBIRS

Using a low temperature reactor, Titan Technologies through its research affiliate Adherent Technologies Inc., is implementing the recycling of plastics and composites. Shredded bottles and B-2 bomber fibre reinforced plastics are turned into high quality liquid chemicals, reusable carbon fibres and other marketable materials, it is claimed. The process is said to be able to process very complex mixtures including photocopiers or computers and municipal plastics wastes without sorting. Further general details are given.

TITAN TECHNOLOGIES INC.; ADHERENT TECHNOLOGIES INC.; US,DEPT.OF DEFENSE

USA

Accession no.609440

Item 215

Antec '96. Volume III. Conference proceedings.

Indianapolis, 5th-10th May 1996, p.3160-4

CONVERSION OF WASTE PLASTICS INTO TRANSPORTATION FUELS

Rangarajan P; Murty M V S; Grulke E A; Bhattacharayya D

(SPE)

Polymers have inherently high hydrocarbon ratios, making liquefaction of waste plastics into liquid fuel feedstocks a potentially viable commercial process. The objective is to characterise the thermal degradation of polymers during hydrogenation. LDPE is studied due to its simple structure. Isothermal and non-isothermal TGA were used to obtain degradation kinetics. Systems of homopolymer, polymer mixtures, and solvent-swollen polymer are studied. The significant variables for liquefaction are pressure, temperature, and time. Product oil viscosity depends on hydrogen pressure. GPC analyses suggest that lower molecular weight polymers degrade at higher rates. Crystallinity increased in the THF and pentanes insolubles as compared to the original LDPE. 15 refs.

USA

Accession no.608720

Item 216

Paper, Film & Foil Converter

70, No.9, Sept.1996, p.86

DUPONT STRENGTHENS POSITION AS TOP FILM PRODUCER

The global activities of DuPont Films are discussed with reference to commercial strategies for profitable growth and by the integration of business and environmental initiatives. The Petretec chemical recycling process for polyester film is briefly described, and details are included of investments.

DUPONT FILMS

WORLD

Accession no.606985

Item 217

Asia-Pacific Chemicals

7, No.7, Sept.1996, p.29/31

LOOKING FOR A GREEN SOLUTION

Gupta N

The Japanese Plastic Waste Management Institute is developing a process to convert PVC and other plastic waste materials to fuel oil through pyrolysis. In Europe, a free market for plastics waste is now being established by the European Plastics Converters over the internet. The company says it will be possible to establish market prices for recyclates at European level on a supply and demand basis. The European market for recycled plastics is currently worth around 1.18bn US dollars and is predicted to reach 2.53bn US dollars by the end of 2001.

WORLD

Accession no.606019

Item 218

Polymer Degradation and Stability

53, No.2, 1996, p.189-97

THERMAL DEGRADATION OF MIXED PLASTIC WASTE TO AROMATICS AND GAS

Kaminsky W; Schlesselmann B; Simon C M

Hamburg,University

A PVC-poor light fraction separated from mixed plastic household waste was pyrolysed to yield aromatic oils and heat-providing gas. Target products were benzene, toluene, xylenes, and styrene. Problematic pollutants were absent or immobilised during pyrolysis in a carbon black matrix. 17 refs.

EUROPEAN COMMUNITY; EUROPEAN UNION; GERMANY; WESTERN EUROPE

Accession no.605034

Item 219

International Polymer Science and Technology

23, No.4, 1996, p.T/102-6

THERMAL METHODS FOR RAW MATERIAL RECYCLING OF PLASTICS

Polaczek J; Machowska Z

Details are given of the thermal methods of recycling plastics. Emphasis is given to pyrolysis, hydrocracking, and gasification. 36 refs.

EASTERN EUROPE; POLAND

Accession no.605015

Item 220

Spartanburg, S.C., 1995, pp.16. 11ins. 15/11/95.

UNITED RESOURCE RECOVERY CORP. LOOK WHO'S TALKING ABOUT US NOW!

United Resource Recovery Corp.

Photocopies of journal articles relating to the Unipet process for the recycling of PETP developed by United Resource Recovery Corp. Details are given of the process which enables contaminated PETP to be recycled by the use of caustic soda which reacts with the PETP to yield ethylene glycol and terephthalic acid, followed by heating and evaporation of the EG which reduces organic impurities to carbon dioxide and water and leaves solid terephthalic salt. Its implications for the industry are also discussed.

USA

Accession no.604408

Item 221

Brussels, c.1995, pp.2. 12ins. 13/10/95.

WASTE TO ENERGY. PLASTICS - RESOURCE OPTIMISATION

APME

Combustion of plastics waste with energy recovery is discussed as one approach to the recycling and waste management of waste plastics. Their role in municipal solid waste combustion is examined, and the importance of refuse derived fuel pellets. Facts supporting the importance of waste to energy projects are reported, and details of some projects currently examining MSW combustion with energy recovery are detailed.

BELGIUM; EUROPEAN COMMUNITY; EUROPEAN UNION; WESTERN EUROPE

Accession no.603963

Item 222

Brussels, c.1995, pp.2. 12ins. 13/10/95.

FEEDSTOCK RECYCLING. PLASTICS - RESOURCE OPTIMISATION

APME

Feedstock recycling is examined as a method of plastics recovery. The range of technologies currently employed are described, and include pyrolysis, hydrogenation, gasification, and chemolysis. Methods for the recycling of mixed plastics wastes are discussed, which include work by BP Chemicals, VEBA Oil, Shell Chemicals and Leunawerke.

BELGIUM; EUROPEAN COMMUNITY; EUROPEAN UNION; WESTERN EUROPE

Accession no.603960

Item 223

Ends Report

No.258, July 1996, p.11

PLASTICS PRODUCERS STRUGGLE TO INCREASE RECOVERY RATES

This article provides information from a report from APME, which shows that recycling and recovery rates have declined since the early 1990's. The report calls for an expansion in incineration - but not recycling - capacity. Brief details are given.

APME

EUROPE-GENERAL

Accession no.603659

Item 224

Polymer Degradation and Stability

53, No.1, 1996, p.129-35

EFFECTS OF ZEOLITES ON THE PYROLYSIS OF POLYPROPYLENE

Zhao W; Hasegawa S; Fujita J; Yoshii F; Sasaki T; Makuuchi K; Sun J; Nishimoto S

Chinese Academy of Sciences; Japan Atomic Energy Research Institute; Kyoto,University

As part of a study of chemical recycling of waste polymers, various types of zeolites were used as catalysts for the pyrolysis of PP and the effects of zeolites on the degradation temp. and pyrolysed products of PP were studied. It was found that the degradation temp. of PP was strongly dependent on the type of zeolite used and the amount added. One type of HY zeolite (320HOA) was shown to be a very effective catalyst. Pyrolysis products, identified by using gas chromatography/mass spectrometry, were also affected by the addition of zeolites. Some of the zeolites did not change the structure of the products, but narrowed the product distribution to a smaller molecule region, while the HY zeolite led to hydrocarbons concentrated at those containing 4-9 carbons. Furthermore, some new compounds with cyclic structures were found in the presence of the HY zeolite. 22 refs.

CHINA; JAPAN

Accession no.600806

Item 225

Popular Plastics and Packaging

40, No.12, Dec.1995, p.57-64

RECYCLING OF POLYMETHYLMETHACRYLATE (PMMA) WASTE

Chandra R; Adab A; Singal A; Rustgi R

Delhi,College of Engineering

This paper reviews recycling technologies of PMMA waste, its applications and its markets. It relates in detail experimentation on thermal and oxidative depolymerisation of PMMA scrap, under nitrogen and oxygen atmospheres, at different heating rates by thermogravimetry and differential scanning calorimetry techniques. 15 refs.

US,FOOD & DRUG ADMINISTRATION

INDIA

Accession no.595438

Item 226

Popular Plastics and Packaging

40, No.12, Dec.1995, p.55-6

PLASTIC RECYCLED INTO RAW MATERIALS

Larane A

ACTIM French Technologies

This article discusses the recycling of plastics back into raw materials, and looks at the recycling activities of several companies across Europe.

BP CHEMICALS; ALPHACAN; ATOCHEM; OTVD; SITA; EREMA; HERBOLD; IFP; ATOHAAS; COMPIEGNE,UNIVERSITE; MICHELIN

AUSTRIA; EUROPE-GENERAL; EUROPEAN COMMUNITY; EUROPEAN UNION; FRANCE; GERMANY; WESTERN EUROPE

Accession no.595437

Item 227

Popular Plastics and Packaging

40, No.7, July 1995, p.51-4

PLASTIC/POLYMER DEGRADATION: A ROUTE OF ECONOMIC UTILISATION AND RECOVERY OF CHEMICALS FROM POLYMER/PLASTIC WASTE

Marathe A B; Nemade S N; Thorat P V

Akola,College of Engineering and Technology

This article highlights the growing problem of plastics waste disposal and then investigates modes of polymer degradation - thermal, chemical, mechanical, photo-degradation, bio-degradation, solvolysis and glycerolysis. 4 refs.

INDIA

Accession no.595430

Item 228
Plastics and Rubber Weekly
No.1645, 19th July 1996, p.7
PACKAGING: THE WASTE CHALLENGE

The UK Department of the Environment has stipulated that each packaging material must achieve an 8% recycling rate by 1998 and 15% by 2001. Pira put the 1993 figure for plastics excluding conversion waste at under 45,000 tonnes or 3%. A company does not have to directly recover its own packaging materials, but will need certificates from an approved recycler or energy recovery scheme showing that the equivalent tonnage of material has been recovered on its behalf. The virgin prices at which a recycled market is sustainable are put at 640-830 pounds sterling, which rules out most of the commonly used packaging plastics.

EUROPEAN COMMUNITY; EUROPEAN UNION; UK; WESTERN EUROPE

Accession no.594965

Item 229
Utech '92. Conference proceedings.
Hague, 31st March-2nd April 1992, p.247-51. 43C6
PROGRESS IN THE TECHNOLOGIES FOR RECYCLING OF PU SCRAPS
Petrone A; Grego S; Chinellato S; Puppin P
ECP Enichem Polimeri
(Crain Communications Ltd.)

It is well known that glycolysis process allows the conversion of PU materials to OH terminated derivatives (polyols). These can be reused, with appropriate formulative modifications, for the preparation of PUs. While work performed up to now is mainly oriented to recycling of rigid and microcellular elastomeric PUs, other important PU types have been rather neglected. One of the objects of this work has been to define glycolysis procedures for other PU materials, and to generate intermediates to be reused also in other fields, thus extending the applicability of the recycled products. A major task of the study has also been to stress the effects of some important process parameters, i.e. nature of PU scraps, type and concentration of glycols, catalysis composition and concentration, on the chemical characteristics of the obtained polyols. Work is focused on the possibility of maximising the amount of PU scraps to be glycolysed and on the criteria for obtaining low OH value polyols. Particular attention is paid to the reduction of by-products of the glycolysis process that can give rise to unfavourable environmental impact. 5 refs.

EUROPEAN COMMUNITY; EUROPEAN UNION; ITALY; WESTERN EUROPE

Accession no.594671

Item 230
European Chemical News (Chemscope)
July 1996, p.28/30
BALANCING ACT KEY FOR SUCCESSFUL RECYCLING
Dhillon P R

Industry experts agree that feedstock recycling is a commercially viable means of recycling waste plastics and forms a significant part of an integrated waste management system. In the Netherlands, the first commercial gasification facility for mixed plastics waste is being developed at the Air Products facility. Startup is planned for 1997 and negotiations are under way for the supply of about 40,000 t/y of plastics waste. In France, a pilot plant for the depolymerisation of PETP will be built by Polyphenix and the EMC Group. A pan-European consortium has a pilot plant based in Grangemouth with a feedstock recycling capacity of about 300 t/y. The technology developed is a fluidised bed thermal cracking process.

WESTERN EUROPE-GENERAL; WESTERN EUROPE

Accession no.593975

Item 231
Converter
33, No.6, June 1996, p.8
CONFLICT OF APPROACH
Darrington R
Orwak Linley Ltd.

This comprehensive article supplies a discussion of the arguments for and against incineration with energy recovery as a technique for treating discarded packaging. The article compares the advantages of this method of dealing with discarded packaging with the standard alternative of salvaging individual materials and recycling them.

EUROPEAN COMMUNITY; EUROPEAN UNION; UK; WESTERN EUROPE

Accession no.591614

Item 232
Shell Chemicals Europe Magazine
No.4, Nov.1995, p.20-3
STOCK OPTION
Mader F; Mennicken T

Feedstock recycling is examined as one solution to Europe's waste management problems. The process is described, and is discussed as one of the options available to recycle plastics, along with mechanical recycling and energy recovery. Statistics are included for production of waste and recovery levels. Germany is examined as an example of leading research and development in feedstock recycling, and details are given of such activities being carried out in the country.

WESTERN EUROPE-GENERAL; WESTERN EUROPE

Accession no.591002

Item 233
Journal of Applied Polymer Science
60, No.10, 6th June 1996, p.1637-44
CONTRIBUTION TO THE MODELLING OF PREDEPOLYMERISATION OF POLYSTYRENE
Swistek M; Nguyen G; Nicole D
Nancy,Universite Henri Poincare

The thermal degradation process of a standard PS having a low polydispersity index was modelled in order to optimise the experimental conditions for the recycling of plastics wastes. The number of initial ruptures, No, in the macromolecular chain during pyrolysis at 350C under nitrogen pressure was determined by GPC results from the experiments with tetralin as the hydrogen donor solvent, assuming that all radicals had been stabilised by this solvent. The calculation showed that there were 23 depropagation reactions, 40 intramolecular transfers and, for n intermolecular transfer, n-2 recombinations and No-n+2 dismutations. 23 refs.

EUROPEAN COMMUNITY; EUROPEAN UNION; FRANCE; WESTERN EUROPE

Accession no.590596

Item 234
Plastics News(USA)
8, No.4, 25th March 1996, p.19
JAPAN'S PROGRAM EMPHASISES LOCALISED RECOVERY EFFORTS
Ford T

A national recycling program passed by the Japanese parliament in June 1995 is underway. Details are given of the plan in which each individual municipality will have virtual autonomy in determining how it wishes to dispose of waste plastic within guidelines, goals and time frames set up by the national government. Under the provisions of the law, recycling programs must be generated for PETP bottles by April 1997, and the municipalities will be able to choose recycling by mechanical means, incineration for energy generation or use of plastic wastes as feedstocks for the manufacture of chemicals and plastics. Statistics for the production of PETP waste are included.

JAPAN

Accession no.589756

Item 235
Plastics News(USA)
8, No.4, 25th March 1996, p.18
MIXED PLASTIC WASTE FUELLING STEEL MANUFACTURERS' OVENS
Ford T

The use of mixed plastics waste as fuel for steel furnaces is briefly examined as NKK Corp., the second largest steelmaker in Japan plans to use about 66 million pounds of it. According to the company's test data, 440 pounds of waste plastics can replace 904 pounds of coke and 198 pounds of pulverised coal in the production of one ton of pig iron. In addition, in Germany, Wirtschaftsvereingung Stahl, a steel industry association, estimates that the German steel industry could use all 1.4 billion pounds of the plastic waste collected by Duales System Deutschland.

NKK CORP.
JAPAN

Accession no.589754

Item 236
Macromolecules
29, No.9, 22nd April 1996, p.3315-6
NOVEL APPROACH FOR THE CHEMICAL RECYCLING OF POLYMERIC MATERIALS: THE NETWORK POLYMER-BIFUNCTIONAL MONOMER REVERSIBLE SYSTEM
Endo T; Suzuki T; Sanda F; Takata T
Tokyo,Institute of Technology

Details are given of the successful construction of a novel reversible system of network polymers between bifunctional monomers by utilising the equilibrium polymerisation system of a spiro orthoester. Molecular structures were determined by NMR and IR spectroscopy. 9 refs.

JAPAN

Accession no.589349

Item 237
Polyurethanes '95. Conference Proceedings.
Chicago, Il., 26th-29th Sept.1995, p.287-90. 43C6
NEW POLYOLS MADE BY GLYCOLYSIS FROM PUR AND PIR RIGID FOAM SCRAP AND THEIR APPLICATIONS
Naber B; Neiss V; Gassan M C
BASF Schwarzheide GmbH
(SPI,Polyurethane Div.)

The application of a glycolysis process with simultaneous deamination to the recovery of polyols from rigid PU and polyisocyanurate foam waste is described. Properties and applications of the polyols obtained are examined.

GETZNER CHEMIE GMBH
AUSTRIA; EUROPEAN COMMUNITY; EUROPEAN UNION; GERMANY; USA; WESTERN EUROPE

Accession no.588963

Item 238
Polyurethanes '95. Conference Proceedings.
Chicago, Il., 26th-29th Sept.1995, p.279-86. 43C6
RECYCLING AND RECYCLED CONTENT FOR POLYURETHANE FOAM
Hicks D A; Hemel C B; Kirk A C; Stapleton R J; Thompson A R
ICI Polyurethanes; ICI Holland BV
(SPI,Polyurethane Div.)

A split phase glycolysis process for the recovery of polyols from PU foam waste is described. Applications of the polyols in the manufacture of flexible and rigid PU foams are examined, and the economics of the process are analysed. 2 refs.

BELGIUM; EUROPEAN COMMUNITY; EUROPEAN UNION; NETHERLANDS; USA; WESTERN EUROPE

Accession no.588962

Item 239

Polyurethanes '95. Conference Proceedings.
Chicago, Il., 26th-29th Sept.1995, p.270-4. 43C6

TERTIARY RECYCLING OF WASTE POLYURETHANE CAR SEAT FOAM

Tatsumoto K; Elam C C; Looker M J; Evans R J
US,National Renewable Energy Laboratory
(SPI,Polyurethane Div.)

The application of a selective pyrolysis process to the recovery of chemicals from waste PU foam is described. The reaction conditions are controlled so that target products can be collected directly from the waste stream in high yields. Molecular beam mass spectrometry is used in small-scale experiments to analyse the reaction products in real time, enabling the effects of process parameters such as temperature, catalysts and co-reagents to be quickly screened. Fixed bed and fluidised bed reactors are used to provide products for conventional chemical analysis to determine material balances and to test the concept under larger scale conditions. Results are presented for the recycling of PU foams from vehicle seats and refrigerators. 12 refs.

USA

Accession no.588960

Item 240

Polyurethanes '95. Conference Proceedings.
Chicago, Il., 26th-29th Sept.1995, p.253-60. 43C6

RECYCLING OF POLYUREA-URETHANE RIM

Kresta J E; Xiao H X; Suthar B; Baeten L; Li X H; Sun S P; Klempner D
Detroit,Mercy University
(SPI,Polyurethane Div.)

Polyurea-urethane reaction injection moulding (RIM) materials were recycled by transesterification catalysed by ethylene glycol (EG). Swelling and Fourier transform IR spectroscopy studies of the reaction products (liquid oligomers and insoluble residues) during the reaction indicated that transesterification proceeded only in the surface layers of the RIM particles. During the reaction the urea and urethane bonds were cleaved, forming liquid oligomers, and all urea groups were transformed into urethane groups. The reaction could be carried out at low EG/RIM ratios, resulting in the elimination of the usual recovery of excess reactant. Adhesives with high lap shear and peel strength were prepared by reacting the liquid oligomers with epoxy resins and blocked isocyanate-terminated PU prepolymers. 14 refs.

USA

Accession no.588958

Item 241

Chimica e l'industria

77, No.2, Feb.1995, Supplt., p.4-6
Italian

CHEMICAL RECYCLING OF PLASTICS

Pilati F
Modena,University

Economic and ecological aspects of chemical recycling are examined, and the application of such processes to the recovery of monomers and intermediates from PETP, polyamides, polyurethanes, polycarbonates, unsaturated polyesters, polyacetals, PMMA and PS is discussed. 17 refs.

SNIA

EUROPEAN COMMUNITY; EUROPEAN UNION; ITALY; WESTERN EUROPE

Accession no.588908

Item 242

Chimica e l'industria

77, No.1, Jan.1995, p.33-6
Italian

USE OF REFUSE DERIVED FUEL ENHANCED WITH POST-CONSUMER PLASTICS FOR THE PRODUCTION OF ELECTRICAL ENERGY BY THE GASIFICATION PROCESS

Barducci G L; Daddi P; Lanzino M; Polzinetti G C; Ulivieri P; Schiona G
SAFI SpA; Replastic

Results are presented of studies undertaken in Italy by SAFI and Replastic of the gasification of refuse derived fuel enriched with post-consumer plastics for the production of electrical energy and gas for use in cement making. 11 refs.

EUROPEAN COMMUNITY; EUROPEAN UNION; ITALY; WESTERN EUROPE

Accession no.586669

Item 243

Kunststoffe Plast Europe

86, No.2, Feb.1996, p.19-21

WEIGHING UP THE ECOBALANCE OF DIFFERENT PLASTICS RECYCLING METHODS

Feuerherd K H
BASF AG

Mechanical recycling, feedstock recycling and thermal energy recovery are all valid methods of recycling plastics. This is the conclusion reached by an ecobalance study coordinated by the TUV Rheinland. According to this study, a mixture of all three recycling methods provides the optimum solution to the problem. The relative use made of each method should be guided by what is technically feasible and cost considerations. 3 refs. Translated from Kunststoffe, 86, No.2, 1996, p.198-201

EUROPEAN COMMUNITY; EUROPEAN UNION; GERMANY; WESTERN EUROPE

Accession no.584783

Item 244
Kunststoffe Plast Europe
84, No.7, July 1994, p.15-6
German; English
PVC FEEDSTOCK RECYCLING
Menges G; Lackner V; Fischer R
RWTH

Recycling of waste PVC is examined with reference to feedstock recovery as an alternative to material recycling, in which additive contents, which are no longer permitted, have an undesirable influence. In order to recycle feedstocks, the PVC has to undergo a dehydrochlorination process. The use of degradative extrusion using a twin-screw extruder and a kneader, is described, and a comparison of the results obtained during dehydrochlorination by these two methods is given. 6 refs.

EUROPEAN COMMUNITY; EUROPEAN UNION; GERMANY; WESTERN EUROPE

Accession no.584594

Item 245
Plastiques Modernes et Elastomeres
46, No.10, Dec.1994, p.26-7
French
CONSORTIUM OF FIVE CHEMICAL COMPANIES STARTS A PILOT PLANT FOR CHEMICAL RECYCLING
Trebord L

Details are given of a pilot plant for the chemical recycling of plastics which has been established at Grangemouth, Scotland, by a consortium consisting of BP Chemicals, DSM, Elf Atochem, EniChem and Petrofina. The plant, capable of treating 300 tonnes of waste yearly, uses a low pressure, fluidised bed thermal cracking process.

BP CHEMICALS LTD.; DSM NV; ELF ATOCHEM SA; ENICHEM SPA; PETROFINA SA

BELGIUM; EUROPEAN COMMUNITY; EUROPEAN UNION; FRANCE; ITALY; NETHERLANDS; UK; WESTERN EUROPE

Accession no.583061

Item 246
High Performance Textiles
Feb.1996, p.12
CARPET WASTE RECYCLING PROJECT LAUNCHED

A pan-European initiative that aims to develop a sustainable closed-loop system for recycling and reuse of materials and energy recovery from post-consumer, as well as post-industrial, carpet waste is being coordinated by DSM and EniChem. Called RECAM, the four major objectives of the project are: to achieve sustainable waste management of post-consumer industrial carpet waste; reduce waste management costs by about 30-50%; obtain a cheap energy supply for industry by generating energy from residues; recover and reuse high quality raw material from post-use carpets, with perhaps 50-60% recovery.

DSM NV; ENICHEM SPA

WESTERN EUROPE-GENERAL; WESTERN EUROPE

Accession no.581631

Item 247
Packaging Week
11, No.35, 22nd Feb.1996, p.18-9
PETP RECYCLING UPDATE
Goddard R

The growth rate for PETP has been one of the highest of all packaging materials, and all indications are that it will continue to be so. Three main factors influence the recycling issue: the economics of collection, the recycling capacity of the industry, and the market for the recovered material and/or products made from it. The latest system for the chemical recovery of PETP has been recently reported by Innovations in PET, an Australian company which claims its RENEW process can tolerate significant quantities of contaminant. The excellent prospects for PETP recycling means that it will be expected to make a large contribution to the EU's 15% overall recycling target.

PETCORE

WESTERN EUROPE-GENERAL; WESTERN EUROPE

Accession no.581623

Item 248
Plastics News(USA)
7, No.28, 11th Sept.1995, p.38
METHOD BREAKS DOWN CHLORINATED PLASTICS
Ford T

Molten Metal Technology Inc. of Waltham, Massachusetts, a company that specialises in the disposal of hazardous wastes, has developed a technology which it claims can break down chlorine-containing plastic waste such as PVC, into harmless components without creating toxic emissions such as dioxin. The article supplies details of the process, which utilises a completely sealed system so there are no remissions. The process, known as Catalytic Extraction Processing, involves emersing the chlorinated plastic in a bath of molten metal, heating to 3,000F and then adding chemical reactants such as lime.

MOLTEN METAL TECHNOLOGY INC.

USA

Accession no.578059

Item 249
Industrial & Engineering Chemistry Research
34, No.12, Dec.1995, p.4514-9
CHEMICAL RECYCLING OF WASTE PS INTO STYRENE OVER SOLID ACIDS AND BASES
Zhang Z; Hirose T; Nishio S; Morioka Y; Azuma N; Ueno A; Ohkita H; Okada M
Shizuoka,University; Toyohashi,University of Technology; Furukawa Electric Co.Ltd.

The catalytic degradation of waste PS into styrene was studied using solid acids and bases as catalysts. Degradation mechanisms are discussed in terms of the depolymerisation reaction. 19 refs.

JAPAN

Accession no.575446

Item 250
Plastics World
53, No.12, Dec.1995, p.25
BIG PLANS FOR CARPET
Schut J H

United Recycling has a novel mechanical process for disassembling and recycling post-consumer carpet. The new patented process actually pulls carpet apart to separate face fibre from backing. The idea is to capture some of the 4 billion lb/year of high-value nylon 6 and nylon 66 that gets thrown away in used carpet. With Fluor Daniel's assistance, United Recycling plans to build a 15-million lb/year plant in 1996 using the new process to recycle post-consumer carpet.

UNITED RECYCLING INC.; FLUOR DANIEL INC.

USA

Accession no.574612

Item 251
Macplas International
Aug.1995, p.64-5
CLOSED LOOP RECYCLING OF HIGH PERFORMANCE ENGINEERING RESINS
Bitritto M M
Hoechst Celanese

Some insight is given into the trade-offs and costs involved in two major closed-loop recycling categories: chemical or thermal depolymerisation to form 'new' starting materials, i.e. monomers that are identical to virgin materials, and direct reuse in which scrap is purified, densified and remelted for moulding, extrusion or other processes. The overall aim will be to integrate recycling into normal business activity.

EUROPEAN COMMUNITY; EUROPEAN UNION; ITALY; WESTERN EUROPE

Accession no.571173

Item 252
Macplas International
Aug.1995, p.61-3
INNOVATIVE SOLUTIONS FOR RECYCLING CAR COMPONENTS
Cortesi P; Petrone A; Cremonini F
ENICHEM

A detailed account is given of current problems and research in recycling of automotive plastics, including physical recycling, chemical recycling, thermal recycling, and design for recycling.

EUROPEAN COMMUNITY; EUROPEAN UNION; ITALY; WESTERN EUROPE

Accession no.571172

Item 253
European Plastics News
22, No.10, Nov.1995, p.26
TO BURN OR NOT TO BURN?
Reade L

A report from APME says that Germany's DSD system should start replacing mechanical recycling with incineration. The report used six environmental criteria, including consumption of resources, global warming potential and acidification, to assess the impact of 12 recycling scenarios. Five of these were by mechanical recycling, five by chemical recycling methods and two by energy recovery. It concluded that collecting, sorting and cleaning plastics packaging was very laborious and should only comprise a small part of all operations. It agreed with an earlier Dutch study which suggested that more than half of plastics packaging should be incinerated for energy recovery.

APME; DSD

EUROPEAN COMMUNITY; EUROPEAN UNION; GERMANY; WESTERN EUROPE

Accession no.570917

Item 254
Modern Plastics International
25, No.11, Nov.1995, p.27
WASTE PACKAGING REPLACES HEAVY OIL IN BLAST FURNACE
Colvin R

The world's first steel mill to replace heavy oil with plastics waste as a reducing agent in manufacturing iron has gone into commercial operations in Bremen, Germany. Stahlwerke Bremen expects to use up to 70,000 t/y of collected household waste from DSD. About 200kg/h plastic pellets are blown into the lower part of the blast furnace where they are cracked into reducing syngas at 2100C. The gas deoxidises the iron ore. The company replaces 30% of the oil previously used with the pellets, which contain approximately 2% PVC. The steel maker is planning a joint venture to collect and separate plastic scrap with lower PVC content to eventually replace oil in its ovens.

STAHLWERKE BREMEN

EUROPEAN COMMUNITY; EUROPEAN UNION; GERMANY; WESTERN EUROPE

Accession no.570889

Item 255

Polymers for Advanced Technologies
6, No.11, Nov.1995, p.688-92

MAKING POLYMER CONCRETE AND POLYMER MORTAR USING SYNTHESISED UNSATURATED POLYESTER RESINS FROM POLY(ETHYLENE TEREPHTHALATE) WASTE

Abdel-Azim A A A; Attia I A
Egyptian Petroleum Research Institute; Cairo,Ain Shams University

Depolymerisation of PETP waste was studied in the presence of manganese acetate catalyst and propylene glycol at different weight ratios. The glycolysed products were analysed for hydroxyl value and the amount of free glycol. A series of unsaturated polyesters based on the glycolysed products, maleic anhydride and styrene, were prepared. Molecular weights and curing behaviour of these polymers were determined. Polymer concrete and polymer mortars made with these resins were investigated for their compressive strength. 13 refs.

EGYPT

Accession no.569361

Item 256

Resources, Conservation & Recycling
15, No.1, Oct.1995, p.51-63

ENVIRONMENTAL IMPACTS BY DISPOSAL OF PLASTIC FROM MUNICIPAL SOLID WASTE

Molgaard C
Denmark,Technical University

An "Ecoprofile" is an assessment of the environmental and resource impacts of a waste disposal process. This paper describes ecoprofiles for six different ways of disposing the plastic fraction in municipal solid waste - two material recycling processes that include separation of the plastic waste, material recycling without separation of the plastic waste, pyrolysis, incineration with heat recovery, and landfill. 17 refs.

DENMARK; EUROPEAN COMMUNITY; EUROPEAN UNION; SCANDINAVIA; WESTERN EUROPE

Accession no.568649

Item 257

INCPEN Journal
No.6, Autumn 1995, p.4

USA AND RECYCLING

Perchard D
Incpen

This article reviews packaging legislation in the USA, and includes details of the laws in several states. It also briefly outlines various packaging waste policies in other parts of the world.

EUROPEAN COMMISSION

EASTERN EUROPE-GENERAL; EUROPE-GENERAL; EUROPEAN COMMUNITY; EUROPEAN UNION; FAR EAST; LATIN AMERICA; UK; USA; WESTERN EUROPE-GENERAL; WESTERN EUROPE

Accession no.568599

Item 258

Polymer Recycling
1, No.3, 1995, p.191-6

RECOVERY OF HYDROCHLORIC ACID FROM THERMAL PROCESSING OF WASTE PVC

Schaub M
Sulzer Chemtech AG

Recycling of plastic materials becomes more important, but it is not possible to make materials of the same quality as virgin materials. For PVC this problem is even more pronounced than for other plastics. For this reason, a process was studied which is able to destroy the waste PVC but which can recover its most important component, chlorine, as a raw material for vinyl chloride monomer manufacture with a very high yield. Most of the energy contained in the PVC can be recovered as electrical power and steam. 4 refs.

SWITZERLAND; WESTERN EUROPE

Accession no.568088

Item 259

Advances in Polymer Technology
14, No.4, Winter 1995, p.337-44

CHEMICAL RECYCLING OF MIXED PLASTICS BY PYROLYSIS

Kaminsky W
Hamburg,University

The use of pyrolysis for the recycling of mixed plastics is discussed and it is shown that fluidised bed pyrolysis is particularly advantageous. It is demonstrated that 25 to 45% of product gas with a high heating value and 30 to 50% of an oil rich in aromatics can be recovered. The oil is found to be comparable with that of a mixture of light benzene and bituminous coal tar. Up to 60% of ethylene and propylene can be produced by using mixed polyolefins as feedstock. It is suggested that, under appropriate conditions, the pyrolysis process could be successful commercially. 23 refs.

EUROPEAN COMMUNITY; EUROPEAN UNION; GERMANY; WESTERN EUROPE

Accession no.566567

Item 260

Angewandte Makromolekulare Chemie
Vol.232, Oct. 1995, p.151-65

PYROLYSIS WITH RESPECT TO RECYCLING OF POLYMERS
Kaminsky W
Hamburg,University

Details are given of the pyrolysis of plastics waste with emphasis given to the use of a heated fluidised bed reactor. Data are given for the pyrolysis conditions of mixed plastics as well as gas composition and high- and low-boiling point fractions. 20 refs.

EUROPEAN COMMUNITY; EUROPEAN UNION; GERMANY; WESTERN EUROPE

Accession no.565722

Item 261
International Polypropylene Conference. Conference Proceedings.
London, 24th-25th Oct.1994, p.165-75. 42C12
RECYCLING OF AUTOMOTIVE PARTS
Glemet M; Buerkle D; Jean A
Appryl; Elf Atochem SA
(Institute of Materials; BASF AG; European Chemical News; Montell Polyolefins)

Plastics, which feature lightweight characteristics, are ideal for heavy-duty automotive applications. The objective is to reduce fuel consumption and carbon dioxide emissions. A variety of plastics is necessary to achieve optimum technical and economic results. It has been found that mechanical recycling is the best recovery option for large PP automotive components, while energy recovery is the solution for most small plastic parts.

EUROPEAN COMMUNITY; EUROPEAN UNION; FRANCE; WESTERN EUROPE

Accession no.564894

Item 262
Plastics High Performance Packaging. Conference Proceedings.
Dusseldorf, 24th-25th May 1994, paper 13. 6P
IS HDPE RECYCLING THE BEST DEAL FOR THE ENVIRONMENT?
O'Neill J
Dow Europe
(SPE,European Sections)

This paper describes a life cycle analysis study conducted to determine the environmental impact of mechanical recycling in comparison to, or in conjunction with, other post-consumer HDPE disposal options - landfill, incineration, and energy recovery. The results are presented and discussed in full.

DSD
EUROPEAN COMMUNITY; EUROPEAN UNION; GERMANY; SWITZERLAND; WESTERN EUROPE

Accession no.564354

Item 263
Hydrocarbon Processing
74, No.5, May 1995, p.109-12
RECYCLE PLASTICS INTO FEEDSTOCKS
Kastner H; Kaminsky W
Hamburg,University

This paper discusses in detail the option of fluidised-bed reactors to crack mixed plastics waste into valuable raw materials, under the headings: thermal cracking for feedstocks, pyrolysis of polyolefins, and other options. 7 refs.

DSM; ELF ATOCHEM; ENICHEM; PETROFINA
EUROPE-GENERAL; EUROPEAN COMMUNITY; EUROPEAN UNION; GERMANY; WESTERN EUROPE; WORLD

Accession no.564254

Item 264
Financial Times
No.32797, 4th Oct.1995, p.14
BURNING ISSUE FOR THE PLASTICS INDUSTRY
Luesby J

A German study funded by plastics producers and recyclers claims that the benefits of incinerating plastics are being seriously overlooked as European Union governments begin imposing recycling targets under the packaging directive. In a life cycle analysis comparing the environmental impact of 12 ways of using spent plastics, the reprocessors claim that the logistics of collecting, sorting and cleaning plastics packaging are prohibitive. The producers are adamant that recycling is the best option for just 15% of plastic packaging, in spite of a German recycling target of 64%.

APME
EUROPEAN COMMUNITY; EUROPEAN UNION; GERMANY; WESTERN EUROPE

Accession no.563240

Item 265
Materials World
3, No.9, Sept.1995, p.426-7
POLYMER COMPOSITES: RECYCLING AND ENERGY RECOVERY
Pickering S; Hornsby P
Nottingham,University; Brunel University

It is reported that growth in the use of thermoset composites is being threatened in some industries by the lack of suitable processes for recycling scrap. Although there are not, at present, large quantities of composites to be recycled, it is essential that recycling processes are developed now, in order to secure the marketplace for thermoset composites in competition with other materials. The University of Nottingham and Brunel University are collaborating in a project to develop new ways of recycling and recovering materials and energy from these materials. Details are given.

EUROPEAN COMMUNITY; EUROPEAN UNION; UK; WESTERN EUROPE

Accession no.562912

Item 266

Polymer Recycling

1, No.2, 1995, p.87-97

RECYCLING GLASS-REINFORCED COMPOSITES. THE VALUE OF GLASS FIBRES

Graham W D

Owens-Corning

Recycling of glass fibre-reinforced plastics is reviewed, with special emphasis on remelting of thermoplastic composites, mechanical recycling of thermoset composites, depolymerisation and dissolution of thermosets and thermoplastics, closed loop recycling of glass, and the use of glass as a mechanical compatibiliser. 32 refs.

USA

Accession no.559825

Item 267

Plastics World

53, No.8, Aug.1995, p.27-8

NEW ALCHEMY FOR PETP ARRIVES

Schut J H

Two new caustic hydrolysis recycling technologies are seeking commercialisation. Each holds a promise of recycling more contaminated streams of PETP less expensively than current methanolysis or glycolysis chemical recycling. An unusual feature of RecoPET's process is a 12 metre high adsorption column, said to remove high levels of pigments and chemicals. One of the features of United Resource Recovery's UnPET technology is a neutralising step which reduces a salt by-product to almost nil. The two recycling technologies are described.

RECOPET; UNITED RESOURCE RECOVERY

EUROPEAN COMMUNITY; EUROPEAN UNION; FRANCE; USA; WESTERN EUROPE

Accession no.559587

Item 268

Antec '94. Conference Proceedings.

San Francisco, Ca., 1st-5th May 1994, Vol.III, p.2939-42. 012

THERMOLYSIS OF POLYETHYLENE

McCaffrey W C; Kamal M R; Cooper D G

McGill University

(SPE)

Experiments were undertaken in the thermolysis of linear LDPE. The process gave a high yield of liquid product, with a significant fraction being converted to unsaturated products which were mainly alpha-olefins suitable for use in the production of synthetic lubricants. 11 refs.

CANADA; USA

Accession no.557735

Item 269

Antec '94. Conference Proceedings.

San Francisco, Ca., 1st-5th May 1994, Vol.III, p.2896-900. 012

REACTIVE EXTRUSION FOR THE HYDROLYTIC DEPOLYMERISATION OF POLYETHYLENE TEREPHTHALATE

Kamal M R; Lai-Fook R A; Yalcinyuva T

McGill University

(SPE)

The hydrolytic depolymerisation of PETP at high pressures and temperatures was studied in a co-rotating twin-screw extruder. Under starve feed operation, using cold or hot saturated water for hydrolysis of molten PETP in the extruder was ineffective, but significant depolymerisation could be achieved using high pressure saturated steam injected only against high back pressures generated in the extruder. The closer the water reactant temperature was to that of the melt on injection, the more effective was the hydrolysis reaction. The reaction could be further improved by optimising the screw speed. 18 refs.

CANADA; USA

Accession no.557727

Item 270

Brussels, 1994, pp.8. 12ins. 24/3/95. 8(13)5

ENERGY RECOVERY THROUGH CO-COMBUSTION OF MIXED PLASTICS WASTE AND MUNICIPAL SOLID WASTE

Mark F E

Dow Chemical Europe

Comprehensive testing programmes have been undertaken by APME's project team at the Wurzburg municipal solid waste combustor plant which have confirmed the beneficial effects of mixed waste plastics in the municipal solid waste energy recovery process. By adding medium and high amounts of polymers to the combustion process, improved burn-out in the gaseous phase and solid residue stage were achieved. In addition, higher polymer contents including PVC did not produce any measurable increase in the presence of dioxins and furans. Higher concentrations of plastics, in fact produced a positive effect on emissions.

APME

BELGIUM; EUROPEAN COMMUNITY; EUROPEAN UNION; WESTERN EUROPE

Accession no.555822

Item 271
Plastics Recycling: Technology Charts the Course. Retec Proceedings.
Schaumburg, Il., 3rd-4th Nov.1994, p.232-45. 8(13)
PYROLYSIS OF POLYMER WASTE
Agarwal K
General Motors Corp.
(SPE,Recycling Div.; SPE,Chicago Section)

Various pyrolysis trials conducted on the General Motors laboratory scale pyrolysis unit are described,as are some trial runs conducted with the SMC Auto Alliance. Several laboratory scale and large-scale pyrolysis trials conducted with thermosets, thermoplastics, paint sludge and auto shredder residue are outlined.

USA

Accession no.553650

Item 272
Plast' 21
No.36, Sept.1994, p.29-30
Spanish
VALUE OF PLASTICS IN ENERGY EXPLOITATION
Linacisoro I

An examination is made of processes used in an incineration plant in Wurzburg, Germany, in which plastics are incinerated together with municipal solid waste to produce electrical and thermal energy. Results are presented of studies of emissions arising from the combustion of wastes containing three different levels of plastics.

APME; WURZBURG,UNIVERSITY; DOW CHEMICAL EUROPE; DUALES SYSTEM DEUTSCHLAND

EUROPEAN COMMUNITY; EUROPEAN UNION; GERMANY; WESTERN EUROPE-GENERAL; WESTERN EUROPE

Accession no.552434

Item 273
European Plastics News
22, No.6, June 1995, p.28-9
IS CHEMICAL RECYCLING THE ANSWER
Williams D

Pending the results of a study on the economics of recycling, Germany is reconsidering ways of dealing with its post consumer packaging waste. It is thought the report will have a serious impact on the future of chemical recycling, and that less waste will be made available for chemical recycling. The implications of this are discussed.

EUROPEAN COMMUNITY; EUROPEAN UNION; GERMANY; WESTERN EUROPE

Accession no.552276

Item 274
Reuse/Recycle
25, No.5, May 1995, p.36-7
ANOTHER ROUTE FOR RECYCLING PET BY DEPOLYMERISATION

The UnPET process for the depolymerisation of PETP has been developed by United Resource Recovery Corp. The process is claimed to efficiently remove impurities present in post consumer PETP scrap, even those present at a concentration of 40% or more. Key stages of the process which is protected by patent, are described, and also details of the company's plans to invest over 5 million US dollars in a new facility based on UnPET technology.

UNITED RESOURCE RECOVERY CORP.

USA

Accession no.552239

Item 275
Patent Number: WO 9424102 A1 19941027
CONVERSION OF NYLON 6 AND/OR NYLON 6,6 TO MONOMERS
Moran E F
DuPont de Nemours E.I.,& Co.Inc.

This is achieved by treatment with aliphatic monocarboxylic acid.

USA

Accession no.549995

Item 276
Shell Chemicals Europe Magazine
No.2, March 1995, p.17-21
BURNING AMBITION
Jones J
Shell Chemicals UK Ltd.

An integrated approach to municipal waste management is put forward as the only sensible solution, selecting from a range of resource management and recovery options. These are examined and discussed, with particular reference to incineration with energy recovery.

EUROPEAN COMMUNITY; EUROPEAN UNION; UK; WESTERN EUROPE

Accession no.549786

Item 277
Plastics News(USA)
7, No.3, 20th March 1995, p.1/21
GERMAN OVERFLOW FLOODS EUROPE
King R

The German recycling industry is examined, amidst fears that more plastic waste is being collected than the country has the capacity to recycle. The case of Beyer Industrieprodukte is mentioned which earned critical media attention, reinforcing the consumers' belief that much of the plastic packaging collected is not recycled. Exports of German collected waste plastics are forcing European neighbours to legislate defensively, it is claimed. Recycling costs in Germany are compared with the US, in particular the costs to the tax payer. Recent German recycling projects are discussed, in particular the use by the Klockner Werker steel mill, which uses plastic granules in place of crude oil in its steel ovens to create chemical reactions, and thereby gets around the German restrictions on plastic incineration.

BEYER INDUSTRIEPRODUKTE GMBH
EUROPEAN COMMUNITY; EUROPEAN UNION; GERMANY; WESTERN EUROPE

Accession no.548395

Item 278
Paper, Film & Foil Converter
69, No.1, Jan.1995, p.61-2
GERMAN COMPANY DEVELOPS RECOVERY CONCEPT FOR PLASTICS
Wolpert V M

This reports on the developments by a German company in the use of gasification in the chemical recycling of plastics waste. Brief details are given.

RHEINBRAUN AG; RWE ENTSORGUNG AG; RWE AG; ENERGIEWERKE SCHWARZE PUMPE AG; RWE GROUP COMPANY RHEINBRAUN AG; RWE-GESSELSCHAFT FUER FORSCHUNG & ENTWICKLUNG MBH; SICOWA; PROKU KUNSTSTOFFVEREDELUNG GMBH
EUROPEAN COMMUNITY; EUROPEAN UNION; GERMANY; WESTERN EUROPE

Accession no.547459

Item 279
Shell Petrochemicals
No.24, 1993, p.28-30
PLASTICS WASTE: IS RECYCLING THE RIGHT ANSWER?
Appelboam V
Shell International Chemical Co.Ltd.

Recycling is often regarded as the ideal means for solving the problems of non-renewable resources and protecting the environment. This article questions the justification for this view and comprehensively assesses the relative merits of various methods of dealing with plastics waste. The article includes the costs of recycling and possible alternatives, citing inclusion of plastics waste in municipal solid waste to energy facilities as being more cost effective.

EUROPEAN COMMUNITY; EUROPEAN UNION; UK; WESTERN EUROPE

Accession no.545634

Item 280
European Chemical News
63, No.1659, 20th-26th March 1995, p.39-40
WATERSHED YEAR FOR GERMAN RECYCLING
Hammond M
Harriman Chemsult

In the second half of 1993 the German government was accused of allowing vast quantities of waste collected by the DSD to be exported. Germany has objected vehemently to the recovery targets set in the recently passed EU directive on Packaging & Packaging Waste as they are lower than those set in Germany's ordinance. To fall in line with the directive, Germany has had to increase its domestic recovery capacity. Consequently, efforts to increase capacity within Germany have progressed rapidly and 1995 promises to be the watershed year in which more plastics packaging waste is recovered in Germany than is exported. Seven companies are currently bargaining over the supply of raw materials for their proposed feedstocks recovery plants. The only contract that has been agreed so far is with Stahlwerke Bremen which has developed a gasification technology to allow it to use plastics waste as a heavy fuel oil substitute.

EUROPEAN COMMUNITY; EUROPEAN UNION; GERMANY; WESTERN EUROPE

Accession no.545424

Item 281
Plastiques Modernes et Elastomeres
46, No.6, July/Aug.1994, p.23
French
CHEMICAL RECYCLING: BASF STARTS UP A PILOT PLANT
Topuz B

An account is given of the chemical recycling activities of BASF in a pilot plant at Ludwigshafen in Germany, where mixed plastics waste is processed to obtain hydrochloric acid, oil, gas, naphtha, aromatics and alpha-olefins.

BASF AG; DUALES SYSTEM DEUTSCHLAND; OTTO GMBH
EUROPEAN COMMUNITY; EUROPEAN UNION; GERMANY; WESTERN EUROPE

Accession no.544261

Item 282
Brussels, c.1995, pp.10. 8x12 ins. 11/1/95. 8(13)5
FUEL FOR THE FUTURE, ENERGY FROM PLASTICS WASTE
APME

The feasibility is discussed of producing energy from household waste. Plastics represent only 7% of all municipal solid waste, but this represents 30% of the energy in waste. Energy recovery systems are explained, and the advantages of such fuel production in terms of reducing the reliance on fossil fuels, imports of coal to Europe, and as a way of dealing with the amounts of domestic waste going to landfill, are examined. Sweden is cited as an example of the efficiency of such a scheme, and the Bollmora integrated energy recovery plant is discussed.

BELGIUM; EUROPEAN COMMUNITY; EUROPEAN UNION; WESTERN EUROPE

Accession no.544029

Item 283

Brussels, c.1994, pp.4. 12ins. 30/9/94. 8(13)5

ENERGY RECOVERY - THROUGH CO-COMBUSTION OF MIXED PLASTICS, DOMESTIC WASTE AND MUNICIPAL SOLID WASTE

APME

Test results are presented and discussed following trials in which energy recovery of mixed plastics domestic waste and municipal solid waste was carried out by means of co-combustion. The research also involved the collection of data relating to emissions, and the levels of halogens, dioxins and furans and heavy metals within the mixed plastics waste.

BELGIUM; EUROPEAN COMMUNITY; EUROPEAN UNION; WESTERN EUROPE

Accession no.544023

Item 284

Packaging Communique

Winter 1994/95, p.2/3

DUTCH STUDY CONFIRMS INTEGRATED APPROACH TO WASTE MANAGEMENT IS WAY FORWARD

This reports on the findings of a new independent waste management study, commissioned by the Dutch Government and the Dutch Plastics Producers. The study compared five recovery models for domestic plastics waste, which included various combinations of mechanical and feedstock recycling, assuming streams for recycling were prepared through different collection routes, and energy recovery.

NETHERLANDS,GOVERNMENT; DUTCH PLASTICS PRODUCERS; CENTRE FOR ENERGY CONSERVATION & ENVIRONMENTAL TECHNOLOGY

EUROPE-GENERAL; EUROPEAN COMMUNITY; EUROPEAN UNION; NETHERLANDS; WESTERN EUROPE

Accession no.542060

Item 285

Packaging Communique

Winter 1994/95, p.1

PLASTICS PACKAGING: LOWERING EMISSIONS IN WASTE COMBUSTION

This provides details from a report by APME called "Energy recovery - through co-combustion of mixed plastics waste and MSW", research data from which provides strong support for energy recovery of lightweight plastics packaging, and indicates the role plastics play in reducing emissions during municipal solid waste combustion.

APME

EUROPEAN COMMUNITY; EUROPEAN UNION; GERMANY; WESTERN EUROPE

Accession no.542057

Item 286

Financial Times

No.32601, 15th February 1995, p.13

PLASTICS WASTE STRIKES OIL

Lindemann M

The waste recycling plant in the Ruhr town of Bottrop has been converting assorted plastics waste into oil for almost a year using a hydrogenation process. The oil is blended in a refinery next to the plant to produce high quality oil-based products. Since last April, the plant has been using 40,000 tonnes of synthetic materials waste collected by the DSD. BASF has built a similar plant at its headquarters in Ludwigshafen. The plant heats the waste to create a mixture of liquids and gases which are then distilled to produce a variety of raw materials such as butane, which can be used to manufacture other chemical products. The company has so far spent DM40m to build a plant that converts 15,000 tonnes of waste.

BASF AG; DSD

EUROPEAN COMMUNITY; EUROPEAN UNION; GERMANY; WESTERN EUROPE

Accession no.541394

Item 287

Plastics and Rubber Asia

9, No.55, Dec.1994, p.22

TOSHIBA TURNS PLASTIC WASTE INTO FUEL

Results from a pilot plant run by Toshiba, which recycles waste thermoplastics into fuel oil, have been sufficiently successful for the company to plan to launch commercial operations in the second half of the fiscal year, commencing in April, it is reported. Different pressures and temperatures during processing release oils with different compositions, and include heavy oil, kerosene and gasoline. Mixed waste can be processed without generating harmful gases, it is claimed.

TOSHIBA CORP.

JAPAN

Accession no.539772

Item 288
Plastiques Flash
Nos.271/2, May/June 1994, p.62-3
French
CHEMICAL RECYCLING OF PETP

An account is given of the Recopet process, jointly developed by Technochim Engineering and Institut Francais du Petrole, for the chemical recycling of PETP. The process, which consists of saponification, purification by chromatographic absorption and acidification, produces a terephthalic acid of high purity.

INSTITUT FRANCAIS DU PETROLE; TECHNOCHIM ENGINEERING; RECOPET PROCESS; VALORPLAST
EUROPEAN COMMUNITY; EUROPEAN UNION; FRANCE; WESTERN EUROPE

Accession no.537883

Item 289
Packaging Week
10, No.29, 15th/22nd/29th Dec.1994, p.4
NEW REPORT PREDICTS A SURGE IN EPS RECYCLING
Pringle D

This provides brief information from a new report, produced on behalf of the British Plastics Federation, by the Centre for Economics & Business Research, which says that over half of the UK's EPS packaging will be reclaimed by the year 2010, using recycling or waste-to-energy schemes.

CENTRE FOR ECONOMICS & BUSINESS RESEARCH; BRITISH PLASTICS FEDERATION
EUROPEAN COMMUNITY; EUROPEAN UNION; UK; WESTERN EUROPE

Accession no.537250

Item 290
Industrial & Engineering Chemistry Research
33, No.12, Dec.1994, p.3004-9
EFFECT OF HETEROGENEOUS SECONDARY PYROLYSIS REACTIONS ON THE THERMAL DECOMPOSITION OF POLYURETHANE SCRAP
Takamoto D Y; Petrich M A
Northwestern University

The beneficial effects are demonstrated of heterogeneous secondary pyrolysis reactions on the liquid products of PU pyrolysis. Pyrolysis volatiles are passed through a packed bed of carbonaceous solids that promote the secondary reactions. Activated carbon and reaction injection moulded PU (RIM) char were found to be suitable bed materials. The long-term object was to develop marketable solid products by pyrolysis of wastes, so obtaining high char yields. In addition to affecting the liquid products, RIM char also increased the total char yield. This result has implications for pyrolysis reactor design. 30 refs.

USA

Accession no.536915

Item 291
Recycle '94. Conference proceedings.
Davos, 14th-18th March 1994, paper 94. 8(13)
MATERIAL OR ENERGY RECOVERY FROM WASTE
Schmitt-Tegge J
Umweltbundesamt Berlin
(Maack Business Services)

A question being disputed in Germany, important with regard to the amendment of the German Waste Act, is which wastes or residuals should or must be recycled, and which can be used for energy recovery? This paper attempts to define the line between material recycling and thermal exploitation. 2 refs.

GERMANY,FEDERAL ENVIRONMENTAL AGENCY
EU; EUROPEAN COMMUNITY; EUROPEAN UNION; GERMANY; WESTERN EUROPE; WESTERN EUROPE-GENERAL

Accession no.535905

Item 292
Recycle '94. Conference proceedings.
Davos, 14th-18th March 1994, paper 86. 8(13)
PROCESSING OF PLASTICS WASTE TO FUEL - BAUFELD'S CONCEPT FOR THE CEMENT INDUSTRY
Ogulin H
Baufeld-Austria GmbH
(Maack Business Services)

The waste management situation in Austria is presented, and it is explained that Baufeld-Austria GmbH has developed a method and concept, with the cooperation of cement plant experts, to enable some Austrian cement factories to responsibly use plastics waste as an energy source. The conditions used for developing the model, relating to fuel quality, environmental protection, and public health, are explained. The Baufeld model for processing of plastics waste is then described. Details of future plans are included.

AUSTRIA; WESTERN EUROPE

Accession no.535897

Item 293
Recycle '94. Conference proceedings.
Davos, 14th-18th March 1994, paper 84. 8(13)
ENERGY RECOVERY FROM MSW COMBUSTORS: THE EFFECT OF POLYMERIC MATERIALS
Mark F E

Dow Chemical
(Maack Business Services)

According to APME, energy recovery should be the preferred waste disposal route for polymeric materials that are very contaminated, bonded, laminated to other materials, or are at the end of their performance with respect to their physical/chemical properties. This paper takes a detailed look at energy recovery from municipal solid waste combustors, and considers the effect of polymeric materials.

EUROPEAN COMMUNITY; EUROPEAN UNION; GERMANY; WESTERN EUROPE

Accession no.535895

Item 294
Recycle '94. Conference proceedings.
Davos, 14th-18th March 1994, paper 83. 8(13)
PACKAGING DERIVED FUEL - A NEW CONCEPT FOR ENERGY RECOVERY
Frankenhaeuser M
Neste Chemicals
(Maack Business Services)

This paper discusses packaging derived fuel (PDF), and aims to provide indications of possibilities for co-combustion of PDF in several European countries. National energy profiles are reviewed in order to evaluate the amount of solid fuel used, then the sourcing and preparation of PDF is evaluated and the cost estimated. The paper then discusses the relevant available combustion technologies, including cost estimates for retrofit. Finally, the market potential for PDF is indicated. 4 refs.

APME; EKONO
EUROPEAN COMMUNITY; EUROPEAN UNION; FINLAND; FRANCE; GERMANY; ITALY; SCANDINAVIA; SWEDEN; UK; WESTERN EUROPE

Accession no.535894

Item 295
Recycle '94. Conference proceedings.
Davos, 14th-18th March 1994, paper 82. 8(13)
MONOCOMBUSTION OF MIXED PLASTICS WASTE
Martin R; Barro C; Beekwilder A; Frankenhaeuser M
Bayer AG; Replastic; Akzo; Neste Chemicals
(Maack Business Services)

This paper demonstrates the technical feasibility of a plastics energy recovery plant using circulating fluidised bed technology from Ahlstrom of Finland. Full details are given of a two-phase test run conducted at Ahlstrom's pilot plant in Karhula, in order to obtain information on the process behaviour when combusting different types of plastics waste. Results are presented and conclusions drawn.

BABCOCK; LURGI; AHLSTROM; EBARA; APME
EUROPEAN COMMUNITY; EUROPEAN UNION; FINLAND; GERMANY; JAPAN; SCANDINAVIA; WESTERN EUROPE

Accession no.535893

Item 296
Recycle '94. Conference proceedings.
Davos, 14th-18th March 1994, paper 81. 8(13)
ENERGY RECOVERY FROM USED PACKAGING
Maijgren B
Duni AB
(Maack Business Services)

This paper reports on the progress of the European Working Group on Energy Recovery (WG4) of the European Committee for Standardisation (CEN). Background information on the CEN is provided. Energy recovery from used packaging is then discussed under these headings: integrated waste and resource management, combustible used packaging as a fuel, fuel characteristics of combustible used packaging, heavy metals, product residues in used packaging, energy recovery from combustible used packaging, conclusions, and work in progress. 3 refs.

EUROPEAN COMMITTEE FOR STANDARDISATION
SCANDINAVIA; SWEDEN; WESTERN EUROPE

Accession no.535892

Item 297
Recycle '94. Conference proceedings.
Davos, 14th-18th March 1994, paper 80. 8(13)
PRAGMATIC APPROACH TO WASTE MANAGEMENT IN THE CITY OF PARIS
Guillet R
Paris,Environment Protection Division
(Maack Business Services)

This paper discusses waste management in the city of Paris. Data on solid wastes in Paris is provided, then the role of incineration, and the use of energy from incineration (heating network, production of electricity, total energy recovery), is examined. Air pollution standards are listed, and a comparison made between the different energies used for a heating network. Finally, future considerations are discussed.

EUROPEAN COMMUNITY; EUROPEAN UNION; FRANCE; WESTERN EUROPE

Accession no.535891

Item 298
Recycle '94. Conference proceedings.
Davos, 14th-18th March 1994, paper 77. 8(13)
PLASTICS AND THEIR CONTRIBUTION TO THE DEVELOPMENT OF ENVIRONMENTALLY FRIENDLY CARS
Buerkle D; Jean A

Elf Atochem SA
(Maack Business Services)

This paper explores the use of plastics in cars to make them more environmentally friendly. It lists major environmental issues. It then discusses in detail: the positive role of plastics during the lifetime of a car (more plastics means less fuel consumption), the fact that automotive plastic parts are user-friendly and safe, the current and future uses of plastics in cars, recovery options for plastics in end-of-life vehicles, mechanical recycling (which is the best recovery option for many large automotive parts), energy recovery (the solution for small plastic parts), and feedstock (or chemical) recycling. Lastly, the way forward is considered.

APPRYL; BP CHEMICALS
EUROPEAN COMMUNITY; EUROPEAN UNION; FRANCE; WESTERN EUROPE

Accession no.535888

Item 299
Recycle '94. Conference proceedings.
Davos, 14th-18th March 1994, paper 71. 8(13)
PRESENT STATUS AND FUTURE OUTLOOK FOR PLASTICS WASTE MANAGEMENT IN JAPAN
Katsumata T
Japan,Plastics Waste Management Institute
(Maack Business Services)

This paper provides a detailed overview of the current plastics waste management situation in Japan. It discusses material, chemical, and thermal recycling, and incineration versus landfill. It also provides a flow sheet showing recycling and the treatment/disposal of plastics waste in Japan in 1991. Conclusions are drawn, and the outlook for the future is considered. 5 refs.

JAPAN

Accession no.535883

Item 300
Recycle '94. Conference proceedings.
Davos, 14th-18th March 1994, paper 69. 8(13)
PLASTICS IN AGRICULTURAL APPLICATIONS: MECHANICAL RECYCLING VS ENERGY RECOVERY
Fernandez A P
ANAIP
(Maack Business Services)

This paper discusses the waste management of agricultural plastics in Spain. Information is provided on the Spanish plastics industry, the use of plastics in agricultural applications, and on the environmental problems thus caused. Solutions are discussed, and details are given on recycling plants in Andalusia, including a process description. Finally, energy recovery from agricultural plastics waste is briefly considered.

EUROPEAN COMMUNITY; EUROPEAN UNION; SPAIN; WESTERN EUROPE

Accession no.535774

Item 301
Recycle '94. Conference proceedings.
Davos, 14th-18th March 1994, paper 56. 8(13)
FEEDSTOCK RECYCLING OF PVC RECOVERY OF HCL BY INCINERATION
Hornig P
Wacker-Chemie GmbH
(Maack Business Services)

This paper focuses on feedstock recycling of PVC, which involves degradation in a full-scale incineration plant. Liberated hydrochloric acid is recovered and used in subsequent stages for the generation of new PVC. Full details are given on the basic concept, plant configuration, the process itself, and the economic aspects involved.

EUROPEAN COMMUNITY; EUROPEAN UNION; GERMANY; WESTERN EUROPE

Accession no.535761

Item 302
Recycle '94. Conference proceedings.
Davos, 14th-18th March 1994, paper 55. 8(13)
CHEMICAL RECYCLING STARTING FROM POST CONSUMER PET WASTE TO POLYMER QUALITY PTA
Benzaria J
Recopet
(Maack Business Services)

This paper describes the Recopet process of chemical recycling for post-consumer PETP waste, which yields polymer-quality purified terephthalic acid. Background information is given on PETP recycling, and two routes of chemical recycling are explained. The Recopet process is described in detail, with information provided on the quality of the products obtained.

INSTITUT FRANCAIS DU PETROLE; TECHNOCHIM ENGINEERING
EUROPE-GENERAL; EUROPEAN COMMUNITY; EUROPEAN UNION; FRANCE; WESTERN EUROPE

Accession no.535760

Item 303
Recycle '94. Conference proceedings.
Davos, 14th-18th March 1994, paper 53. 8(13)
RAW MATERIAL RECYCLING - A SOLUTION FOR PLASTICS WASTE
Troussier C
BP Chemicals
(Maack Business Services)

This paper analyses the difficulties of the two traditional plastics waste recovery routes, energy recovery and mechanical recycling, and goes on to introduce a third method - raw material recycling (also called feedstock recycling). This is explained in detail and an example is given of how it could work in a typical European city.

PLASTICS TO FEEDSTOCK RECYCLING CONSORTIUM; DSM; ELF ATOCHEM; PETROFINA; ENICHEM

EUROPE-GENERAL; EUROPEAN COMMUNITY; EUROPEAN UNION; FRANCE; GERMANY; SWITZERLAND; UK; WESTERN EUROPE

Accession no.535758

Item 304

Recycle '94. Conference proceedings.
Davos, 14th-18th March 1994, paper 51. 8(13)

OPTIONS FOR PRIMARY RECYCLING OF PLASTIC RECYCLABLES

Gebauer M
Leuna-Werke AG
(Maack Business Services)

This paper explores the options available for the primary recycling of plastics. It considers the problems of chemical recycling, preparatory treatment (volume reduction, slurry, molecular weight reduction), recycling methods and concepts (primary recycling of pre-treated mixed plastic recyclables in a petroleum refinery, combined hydrogenation/cracking, gasification procedures, and modular concepts for recycling centres), and finally, practical results obtained at Leuna-Werke AG (decompositional extrusion, steam cracker, and visbreaking trials).

EUROPEAN COMMUNITY; EUROPEAN UNION; GERMANY; WESTERN EUROPE

Accession no.535756

Item 305

Recoup Data Digest

No.6, Nov.1994, p.3

RECOUP TAKES STOCK OF DEVELOPING TECHNOLOGY

Details are given of a visit by RECOUP to BP Chemical's feedstock recycling demonstration unit in Sunbury. The feedstock recycling technology has been developed by a consortium of companies, and will enable polyolefin rich plastic waste from domestic and commercial sources to be vapourised and then condensed to form a hydrocarbon wax. This can then be used to feed existing petrochemical crackers to produce polymers indistinguishable from virgin material, it is claimed.

RECOUP

EUROPEAN COMMUNITY; EUROPEAN UNION; UK; WESTERN EUROPE

Accession no.534464

Item 306

Materials Recycling Week

164, No.15, 9th Dec.1994, p.10/14

THE HEAT IS ON FOR ENERGY FROM WASTE

Ray A

Controversy surrounding thermal recycling as a means of recovering energy from waste plastics is reported in the light of an EPA report which claims that there are no safe levels of dioxins which are sometimes produced during incineration. The case for energy from waste is put forward as an alternative to landfilling or mechanical recycling of contaminated and comingled waste.

EUROPEAN COMMUNITY; EUROPEAN UNION; UK; WESTERN EUROPE

Accession no.534444

Item 307

Journal of Coated Fabrics

Vol.23, April 1994, p.274-9

RECYCLING OF PVC-COATED FABRICS

Saffert R
Solvay Kunststoffe GmbH

Techniques for the material recycling of PVC-coated PETP tarpaulins are considered with reference to grinding at room temp., grinding at low temps. and processes using solvents. It is shown that combined reprocessing of the two polymers is not viable and that the same applies for solvent separation. The use of an incineration process to generate energy and permit recovery of chlorine from the PVC is discussed. (Techtextil Symposium 94, Frankfurt am Main, Germany, June 1994)

EUROPEAN COMMUNITY; EUROPEAN UNION; GERMANY; WESTERN EUROPE

Accession no.534413

Item 308

European Plastics News

21, No.10, Nov.1994, p.18-9

PAN-EUROPEAN RECYCLING CONSORTIUM OPENS PILOT PLANT

A consortium of five European polymer producers opened a new chemical recycling pilot plant at BP Chemicals's Grangemouth facility in October. The 100 kg/hr plant uses a patented fluidised bed thermal cracking process to transform mixed waste plastic into a wax which can be used a a substitute petrochemicals feedstock in existing crackers. The fluidised bed technology is suitable for scale up to comparatively small commercial capacities of between 25,000 and 100,000 t/y. The aim of the new Grangemouth plant is to further assess the technology, evaluate operating costs and identify the investment needed for commercial operation.

BP CHEMICALS LTD.

EUROPEAN COMMUNITY; EUROPEAN UNION; UK; WESTERN EUROPE

Accession no.533656

Item 309
Plastics and Rubber Weekly
No.1560, 4th Nov.1994, p.6
CHEMISTRY FEEDS RECYCLING HOPES

Following last month's launch of the feedstock recycling plant at BP Chemicals Grangemouth, the consortium partners - BP Chemicals, DSM, Elf Atochem, Enichem and Petrofina - claim that semi-commercial start-ups using their new polymer cracking technology could be in place by the year 2000 if the right investment were forthcoming. Ultimately, at 300 tonnes a year, the aim is to commercialise a process which generates a product for which there is a guaranteed end market. The emerging process relies on thermal cracking of shredded plastics in a bed fluidised by hydrocarbon gas. The final waxy product has responded well as a feedstock when tested in a pilot steam cracking plant at DSM and in a fluidised catalytic cracker unit at Fina.

BP CHEMICALS LTD.
EUROPEAN COMMUNITY; EUROPEAN UNION; UK; WESTERN EUROPE

Accession no.533620

Item 310
New Materials/Japan
Dec.1994, p.15
NON-FLAMMABLE WASTE PLASTIC OILIFICATION PROJECT TO START

MITI is currently undergoing research to develop technology for the recycling of non-flammable plastics such as those used in business machines and computers. Their National Institute for Resources and Environment plans to decompose, without the production of harmful substances, non-flammable polymers by means of liquid phase hydrocracking, and to recover from them light oils such as benzene, toluene and xylene. The key to the technology, it is claimed, lies in the development of a catalyst which will be able to combine hazardous substances such as bromine and chlorine contained in the waste plastics.

JAPAN,MINISTRY OF INTERATIONAL TRADE AND INDUSTRY
JAPAN

Accession no.532783

Item 311
Chemical Week
155, No.15, 26th Oct.1994, p.22
EUROPEANS SHARE EXPERTISE TO CRACK POLYMER RECYCLING
Wiesmann G

A consortium of five European plastics manufacturers has developed a process for plastics-to-feedstock recycling that could prove cheaper and more versatile than existing processes. A 750 m.t./year pilot plant has been recently commissioned at BP Chemical's Grangemouth, Scotland, site. At the heart of the low-temperature pyrolysis process is a fluidised bed of sand into which waste plastic is introduced and in which impurities such as metal and paper are trapped. BP Chemicals is confident that the subsequent product will be suitable for petrochemical or refinery applications. The consortium estimates that plants of 25,000 m.t./year would be commercially viable at gate fees of DM300/m.t., with total operating costs estimated at DM450/m.t.

WESTERN EUROPE; WESTERN EUROPE-GENERAL

Accession no.532357

Item 312
European Chemical News
62, No.1640, 24th Oct.1994, p.43
BP CONSORTIUM SCALES UP FEEDSTOCK RECYCLING

The BP-led feedstock recycling consortium recently unveiled its new larger-scale fluidised bed pyrolysis pilot plant, located on the BP refinery site at Grangemouth. The 2 tonne/day plant will take mixed plastics waste from a variety of sources to provide more extensive trial results, to be used in the conceptual design of a 25,000 t/y semi-commercial demonstration plant. The consortium envisages a series of plants, of around 25,000-50,000 t/y, scattered across Europe.

BP CHEMICALS LTD.
EUROPEAN COMMUNITY; EUROPEAN UNION; UK; WESTERN EUROPE

Accession no.531541

Item 313
Financial Times
No.32513, 2nd Nov.1994, p.18
PLACE FOR PLASTICS
Lascelles D

A 20ft-high mini plant has been built at BP's refinery at Grangemouth, near Edinburgh, as an experiment by several leading European petrochemical companies to address the difficult issue of plastic recycling. The 750 t/y pilot plant accepts mixed plastics ground into pieces a maximum of 2cm across and passes them over hot sand which converts them into a gas. This is distilled back into plastic feedstock which can be fed back into the petrochemical plant to make fresh plastic. This process could be replicated at many small plants, located at chemical works or even beside municipal waste tips.

BP CHEMICALS LTD.
EUROPEAN COMMUNITY; EUROPEAN UNION; UK; WESTERN EUROPE

Accession no.531531

Item 314
Food, Cosmetics & Drug Packaging
Nov.1994, p.7
SINGLE-STEP PROCESS CONVERTS PLASTICS TO FUEL

This reports on a process developed at the University of Tokyo to convert waste polyolefin plastic into a mixture of high quality fuels. Details of the process, which uses a carbon catalyst, are given.

TOKYO,UNIVERSITY
JAPAN
Accession no.531423

Item 315
Tokyo, c.1994, pp.4. 12ins. 6/6/94. 242C21-6124-8(13)
FOAMED STYROL LIQUEFACTION TREATMENT MACHINE MODEL FL-50/FL-100
Anchorman Corp.

The use is described of a foamed Styrol liquefaction treatment machine which has been developed as an environmentally acceptable method of recovery of waste foamed styrene.Details are given of the liquefaction treatment which consists of four processes: crushing and removal of foreign substances; heating, gasification and pyrolysis; cooling and liquefaction; and the recycling of resultant liquid as solvent for use in the first three processes.

JAPAN
Accession no.530620

Item 316
Plastics Recycling Update
7, No.10, Oct. 1994, p.5-6
CHEMICAL RECYCLING NEWS

Three recycling news items are very briefly reported upon: a Canadian-developed pyrolysis technology that converts plastics scrap into alpha-olefins, a scrap-plastics-to-monomers system under construction in Scotland, and statistical forecasts on chemical recycling in Germany for 1996.

CANADA,CENTRE FOR MINERAL & ENERGY TECHNOLOGY; BRITISH PETROLEUM; DSM; ELF ATOCHEM; ENICHEM; PETROFINA; BASF; RWE; VEBA
CANADA; EUROPEAN COMMUNITY; EUROPEAN UNION; GERMANY; UK; WESTERN EUROPE
Accession no.529820

Item 317
Communique
Aug.1994, p.4
LATEST RESEARCH STRONGLY SUPPORTS ENERGY RECOVERY

This reports on the results of a research project evaluating the potential for refuse derived fuel and packaging derived fuel. Full details of the research and its findings are provided.

APME; FINLAND,GOVERNMENT
EUROPE-GENERAL; EUROPEAN COMMUNITY; EUROPEAN UNION; FINLAND; GERMANY; SCANDINAVIA; WESTERN EUROPE
Accession no.529801

Item 318
Packaging Week
10, No.22, 27th Oct.1994, p.19
SWEDISH WASTE PLAN ENTERS SECOND PHASE

This article details the second phase of the Swedish Government's comprehensive strategy for dealing with packaging waste, and explains that paper, board, plastics, steel and aluminium have been added to the list of packaging materials that must now be recovered by reuse, recycling, and incineration with energy recovery.

SWEDEN,GOVERNMENT; SWEDEN,NATIONAL ENVIRONMENT PROTECTION BOARD
SCANDINAVIA; SWEDEN; WESTERN EUROPE
Accession no.529615

Item 319
Materials Recycling Week
164, No.9, 28th Oct.1994, p.4
PLASTICS INCINERATION CHEAPEST RECYCLING, SAY DUTCH

This article provides brief information from a study carried out in the Netherlands, which revealed that combustion with energy recovery is the most economical way of processing waste plastics, whilst mechanical recycling is the most environmentally attractive option.

NETHERLANDS,GOVERNMENT
EUROPEAN COMMUNITY; EUROPEAN UNION; NETHERLANDS; WESTERN EUROPE
Accession no.529612

Item 320
Antec '93. Conference Proceedings.
New Orleans, La., 9th-13th May 1993, Vol.III, p.3195-7. 012
EMERGING TECHNOLOGIES IN PLASTICS RECYCLING
Meszaros M W
Amoco Chemical Co.
(SPE)

Techniques for the chemical recycling of plastics into monomers and petrochemical feedstocks are described, including chemical and thermal depolymerisation, pyrolytic liquefaction, pyrolytic gasification and partial oxidation.

BRITISH PETROLEUM CO.PLC
EUROPEAN COMMUNITY; EUROPEAN UNION; UK; USA; WESTERN EUROPE
Accession no.528879

Item 321
Antec '93. Conference Proceedings.
New Orleans, La., 9th-13th May 1993, Vol.III, p.3192-4. 012
COMPARISON OF PETP HYDROLYSIS AND GLYCOLYSIS RATES AT HIGH TEMPERATURES
Campanelli J R; Kamal M R; Cooper G D; Brues M
McGill University
(SPE)

The glycolysis of PETP was studied in a batch reactor at 265C. The reaction extent in the initial period was determined as a function of reaction time using a thermogravimetric technique. The rate data were shown to fit a second order kinetic model at small reaction times. An initial glycolysis rate was calculated from the model and was found to be over four times greater than the initial rate of hydrolysis under the same reaction conditions. 4 refs.

CANADA; USA

Accession no.528878

Item 322
Plastics News(USA)
6, No.24, 15th Aug.1994, p.12
ORE. JUDGE WON'T DECLARE PYROLYSIS RECYCLING
Gardner J

An Oregon judge is reported to have put up another obstacle in the way of the plastics industry's attempt to designate pyrolysis of some plastics as recycling. The SPI's request for a judgement declaring that the industry's chemical recycling project in Washington state should be counted as recycling, even when the end product is used for fuel, has been denied. Details are given.

SPI
USA

Accession no.528240

Item 323
Automotive Engineering
102, No.8, Aug.1994, p.29-31
AUTOMOTIVE SHREDDER RESIDUE: THREE RECOVERY CHOICES

This article examines the recovery of automobile shredder residue (ASR), and considers three recovery options: ASR as a landfill day cover, ASR in the production of composite materials, and the pyrolysis of ASR to recover chemical feedstock. All are discussed in detail.

CANADA,NATIONAL RESEARCH COUNCIL
CANADA

Accession no.528129

Item 324
Kautchuk und Gummi Kunststoffe
47, No.8, Aug.1994, p.596-601
German
PYROLYSIS, HYDROGENATION, GASIFICATION
Aurich H P
Huels AG

Processes for feedstock recycling developed since the early 1970s are discussed. The recovery of plastics as synthetic crude oil by thermal pretreatment, i.e. the degradation of polymer molecules into gaseous and oily constituents, is described. The installation of collection and sorting systems for used plastics is considered and the introduction of a 'gate fee', which has to be charged for bringing such systems into operation, is outlined. 19 refs.

EUROPEAN COMMUNITY; EUROPEAN UNION; GERMANY; WESTERN EUROPE

Accession no.528053

Item 325
Patent Number: US 5312898 A 19940517
PROCESS FOR DEPOLYMERISATION OF PAN INTO WATER-SOLUBLE BY-PRODUCTS, INCLUDING AMMONIA AND LOW MOLEC.WT. CARBONACEOUS MATERIALS, USING HOT WATER
Siskin M; Saleh R Y; Knudsen G A
Exxon Research & Engng.Co.

This process is carried out at a temp. from about 200C up to the critical temperature of water at autogenous pressure. PAN is degraded without the production of toxic hydrogen cyanide as a by-product.

USA

Accession no.526124

Item 326
Brussels, 1994, pp.6. 12ins. 25/5/94. 42C11-6P21-8(13)
WEIGHING UP THE OPTIONS: A COMPARATIVE STUDY OF RECOVERY AND DISPOSAL ROUTES. SUMMARY REPORT
APME

HDPE in the solid waste stream was examined from the view of finding the optimum method of recycling with reference to waste disposal options, energy conservation and environmental impacts. Four primary options were investigated: mechanical recycling; incineration with energy recovery; incineration, and landfill, using a life cycle analysis. It was concluded that in the best case scenario examined, the most environmentally efficient HDPE disposal system from an energy and resource optimisation standpoint is a combination of mechanical recycling and energy recovery.

BELGIUM; EUROPEAN COMMUNITY; EUROPEAN UNION; WESTERN EUROPE

Accession no.525731

Item 327
Financial Times
No.32459, 31st Aug.1994, p.12
SIMPLY TAKE ONE WASHING MACHINE
Terazono E

Demand for plastics recycling technology has risen in the past few years in Japan and electronics makers face increasing pressure from the government to dispose of their own products discarded by consumers. In 1990, Toshiba initiated a development programme for recycling chloride plastics, which account for 25% of all plastics. Researchers discovered that adding a high-density alkaline solution when heating the chloride plastics turned the hydrogen chloride into a harmless salt. The new technology can be applied to more than 90% of all plastics, the proportion that decomposes when heated. The prototype system produces more than 200 litres of fuel out of 250kg of plastics in 11 hours.

TOSHIBA CORP.
JAPAN

Accession no.525000

Item 328
Paper, Film & Foil Converter
68, No.7, July 1994, p.63/4
THERMAL RECYCLING OF PLASTICS GAINING POPULARITY IN JAPAN
Wolpert V M

Thermal recycling of plastics is becoming a more popular option in Japan, largely because of a lack of landfill sites, and also because of the materials' potential as an untapped source of energy. The article supplies brief details of the advantages of thermal recycling.

JAPAN,PLASTICS WASTE MANAGEMENT INSTITUTE
JAPAN

Accession no.524951

Item 329
Modern Plastics International
24, No.8, Aug.1994, p.20/2
AUTO PARTS ARE DISSOLVED IN NEW SYSTEM FOR RECYCLING
Mapleston P

Selective dissolution of multi-material automobile components could prove a cost-effective alternative to mechanical, thermal and chemical post-consumer recycling technologies. German company Wietek is already running lab-scale trials at a plant in Saarbrucken and a small-scale commercial plant at Nohfelden will begin operating at the end of this year. The technology dissolves the plastics at high rates, around 10 min, and then reprecipitates them in pure form, enabling them to be reused in exactly the same applications. Typical solvents are ketones and acetic acid esters. The bulk of Wietek's work has been done on styrene-maleic anhydride. A further application is in recovery of PVC from wiring harnesses.

WIETEK
EUROPEAN COMMUNITY; EUROPEAN UNION; GERMANY; WESTERN EUROPE

Accession no.524422

Item 330
Polymer Recycling
1, No.1, 1994, p.3-11
RECYCLING AND ECOLOGY
Thalmann W R
TS Oeko-Engineering AG

This paper focuses on key energy-related waste management issues, namely: processing and waste, criteria for assessing recycling methods and their ecological merit, and the ecological benefits of recycling. Recycling targets and the consumption of resources are also discussed. Finally, this paper looks at the recyclable quantities, the expectations, and the economy in plastics recycling. 13 refs.

DSD; BUWAL
EUROPEAN COMMUNITY; EUROPEAN UNION; GERMANY; SWITZERLAND; WESTERN EUROPE

Accession no.524083

Item 331
Materials Reclamation Weekly
163, No.22, 29th July 1994, p.6
GOVERNMENT GIVES THUMBS UP FOR WASTE-TO-ENERGY

This article discusses the UK Government's plan to make incineration with energy recovery play a larger role in waste management, following a report from the Royal Commission on Environmental Pollution. Recommendations from the report are included.

UK,GOVERNMENT; UK,ROYAL COMMISSION ON ENVIRONMENTAL POLLUTION; PRODUCER RESPONSIBILITY INDUSTRY GROUP
EUROPEAN COMMUNITY; EUROPEAN UNION; UK; WESTERN EUROPE

Accession no.523213

Item 332
Plastics and Rubber Weekly
No.1545, 22nd July 1994, p.7
CRACKING SOLUTION IS NEAR

This article examines the progress being made in methods of converting plastics into chemical feedstocks. BASF is setting an ambitious pace with its feedstock recycling programme with a 1996 target date for a 300,000 t/y plant to be fed with waste plastics from the DSD/DKR system. The process uses a confidential catalyst system and is described as similar to pyrolysis. A 15,000 t/y pilot plant started up in April. Another German approach, the 40,000 t/y hydrogenation plant operated by Veba Oel and Ruhrkohle Umwelt, was opened in April. In the UK, a feedstock recycling pilot plant at Grangemouth employing fluidised bed technology developed by BP Chemicals is scheduled for start-up later this year.

BASF AG; BP CHEMICALS LTD.; VEBA OEL AG; RUHRKOHLE AG
EUROPEAN COMMUNITY; EUROPEAN UNION; GERMANY; UK; WESTERN EUROPE
Accession no.522964

Item 333
Chemical Week
155, No.3, 27th July 1994, p.41
BASF TAKES A CHANCE ON CARPET RECYCLING
Fairley P

BASF Corp. announced earlier this year a commitment to recycle carpets made after February 1 with its nylon 6 fibre. The company claims the carpets will be an economical source of caprolactam monomer for making new nylon 6 fibre. Last year BASF started up a 1.3 million lbs/year pilot nylon separation plant in Cambridge, ON. Experts say there is no guarantee that the recycling of post-consumer carpeting will ever compete economically with virgin materials. However, with waste-to-energy growing more expensive and less politically acceptable, much attention has focused on depolymerisation.

BASF CORP.
CANADA; USA
Accession no.522917

Item 334
Kunststoff Journal
28, No.2, April 1994, p.34-5
German
AVOID, DECREASE, RE-USE: MECHANICAL AND CHEMICAL PROCESSING

Du Pont de Nemours has developed strategy of 'avoid decrease, re-use' in relation to plastic products. Packaging is avoided altogether in some instances, e.g. transport of chemicals and plastic granules by tanker. A number of mechanical recycling methods are described. Chemical recycling methods may either concentrate on specific raw materials, e.g. glycolysis, or handle mixed plastic waste, e.g. hydrogenation. Du Pont has two factories where the intermediate products of PETP are processed by glycolysis and new PETP is obtained.

DU PONT DE NEMOURS GMBH
EUROPEAN COMMUNITY; EUROPEAN UNION; GERMANY; WESTERN EUROPE
Accession no.522714

Item 335
European Plastics News
21, No.7, July/Aug.1994, p.10
STUDIES BACK USING WASTE PACKAGING AS A FUEL

Two new studies have demonstrated the significant benefits of using waste packaging as a substitute fuel. The first study, a joint project between the Finnish Government and manufacturers of plastics and packaging, supported by APME, examined the environmental impact of burning packaging derived fuel (PDF), which consists of separated and shredded paper and plastics and has a calorific value of approximately 20 MJ/kg. The second study, led by APME, concentrated on the potential for PDF-based energy generation across Europe. According to the studies, combustible used packaging can be converted into a fuel with a higher energy value than peat, wood or brown coal. The amount of fuel that could be saved each year by using PDF is estimated to be equivalent to 14 million tonnes of oil.

APME
FINLAND; SCANDINAVIA; WESTERN EUROPE
Accession no.522101

Item 336
Macplas
19, No.157, April 1994, p.107-8
Italian
SEPARATING PLASTICS FROM SOLID WASTES
Fiore L; Vezzoli A
CSI Montedison

Results are presented of a study undertaken by CSI Montedison and Milani Resine of the separation of individual polymers from mixed plastics waste using gravimetric and solvent techniques.

MILANI RESINE SPA
EUROPEAN COMMUNITY; EUROPEAN UNION; ITALY; WESTERN EUROPE
Accession no.521712

Item 337
Food, Cosmetics & Drug Packaging
July 1994, p.2
COMMERCIAL SYSTEM TURNS WASTE PLASTIC INTO OIL

It is reported that Hitachi Zosen Corp. of Japan has begun trial operation of a commercial-scale device for the conversion of waste plastic to oil. Details of the device are provided. It is also reported that Mitsubishi Heavy Industries Ltd. (MHI) has installed a pilot plant for producing gas fuel from waste plastics. The MHI system is described.

HITACHI ZOSEN CORP.; MITSUBISHI HEAVY INDUSTRIES LTD.; KANSAI ELECTRIC POWER

CO.INC.; CHUGOKU ELECTRIC POWER CO.INC.; JAPAN,MINISTRY OF INTERNATIONAL TRADE & INDUSTRY
JAPAN
Accession no.520979

Item 338
New Scientist
143, No.1932, 2nd July 1994, p.20
CAN BURNING PLASTICS BE GOOD FOR THE WORLD?
Coghlan A

This article reports on a full-scale study of the incineration of plastics with ordinary municipal waste, conducted in Germany by APME. The results, which indicate that burning waste plastics does less damage to the environment than other waste disposal methods, are commented upon by the plastics industry, and environmentalists.

ASSOCIATION OF PLASTICS MFRS.IN EUROPE; FRIENDS OF THE EARTH; DOW CHEMICAL EUROPE
EUROPE-GENERAL; EUROPEAN COMMUNITY; EUROPEAN UNION; GERMANY; WESTERN EUROPE
Accession no.520976

Item 339
Materials Reclamation Weekly
163, No.20, 15th July 1994, p.8
GERMANY BURNS RECYCLING TRAIL

A successful pilot waste-to-energy scheme, developed by Siemens AG of Germany, is to go into normal use, it is reported. Details of the combustion process, which generates electricity and yields recyclable materials, are given.

SIEMENS AG
EUROPEAN COMMUNITY; EUROPEAN UNION; GERMANY; WESTERN EUROPE
Accession no.520972

Item 340
Plastics and Rubber Weekly
No.1542, 1st July 1994, p.7
POLYMERS FOR POWER

Detailed co-combustion trails at the Wurzburg energy recovery plant in Germany have confirmed the positive beneficial effects of mixed waste plastics in the municipal waste energy recovery process, according to the leader of the project sponsored by APME. Research has shown that plastics' contribution to higher heat combustion levels ensures lower carbon monoxide emissions and sulphur dioxide concentrations are reduced. The tests showed that dioxin and furan emissions remained comfortably below new and stringent German standards. It is also claimed that greater burn out during the combustion phase minimises potential leachates in ash residues so that these have a greater use in secondary applications such as road construction and landfill coverage.

ASSOCIATION OF PLASTICS MFRS.IN EUROPE
EUROPEAN COMMUNITY; EUROPEAN UNION; GERMANY; WESTERN EUROPE
Accession no.519784

Item 341
European Chemical News
61, No.1623, 20th June 1994, p.27
PLASTICS IMPROVES MSW INCINERATION, SAYS APME

Independently monitored trials at a commercial municipal solid waste (MSW) incinerator and energy recovery plant in Wurzburg, Germany, have revealed that the addition of mixed plastics wastes, including PVC, to MSW has no adverse effect on incinerator emissions. APME claims plastics contribute to higher heat combustion levels, improve the energy yield, reduce emissions of carbon monoxide and sulphur dioxide and, most significantly, lead to no increase in dioxin and furan emissions. APME says that while co-combustion should play an important role for plastics wastes that are highly contaminated, bonded with other materials or degraded, it should be treated equally with other recovery means.

APME
EUROPEAN COMMUNITY; EUROPEAN UNION; GERMANY; WESTERN EUROPE
Accession no.519520

Item 342
Chemistry & Industry
No.10, 16th May 1994, p.366
FROM DUSTBIN TO DYNAMO
Nathan S

This article considers the problems of disposing of increasing amounts of packaging waste in Europe. Increasingly stringent regulations are closing off the option of landfills, and feedstock recycling schemes cannot yet cope with large quantities of waste. The article focusses on a third option: that of incineration with energy recovery.

APME
WESTERN EUROPE; WESTERN EUROPE-GENERAL
Accession no.516128

Item 343
Journal of Applied Polymer Science
52, No.9, 31st May 1994, p.1353-5
HYDROLYSIS OF WASTE PETP BY SULPHURIC ACID AT 150 C FOR A CHEMICAL RECYCLING
Yoshioka T; Sato T; Okuwaki A
Tohoku,University

Waste PETP powder was hydrolysed to terephthalic acid and ethylene glycol in relatively dilute sulphuric acid and the sulphuric acid can be reused by recovery methods such as dialysis. 8 refs.

JAPAN

Accession no.513685

Item 344

Japan Chemical Week

35, No.1775, 19th May 1994, p.4

PROMOTING PLASTICS RECYCLING

The Japanese petrochemical industry is working on methods of plastics waste recycling, in cooperation with the Plastic Waste Management Institute. A method of thermally and chemically decomposing plastics waste into gas, oil and raw materials is presently being developed. Japan, where incineration as a waste disposal method has long been in common use, shows the highest incineration rate, 72% in 1986, in comparison with other countries with a higher rate of landfill disposal. Japan's technology for the treatment of flue gas generated in waste incineration is among the most advanced in the world.

JAPAN,PLASTICS WASTE MANAGEMENT INSTITUTE

JAPAN

Accession no.513368

Item 345

Journal of Thermoplastic Composite Materials

7, No.1, Jan.1994, p.64-74

NEW DEVELOPMENTS IN CHEMICAL RECYCLING AS A SINK FOR PROBLEMATIC WASTE FROM FIBRE-REINFORCED PLASTICS

Menges G

Institut fuer Kunststoffverarbeitung

Three alternative techniques for waste disposal are described which involve reprocessing routes which can be conducted in self-contained plants without any emissions. The techniques are the Thermoselect process used in Northern Italy (Verbania), the high temp. shaft furnace process (Voest) and the high temp. combustion plant using pure oxygen and gasification (University of Aachen). The suitability of the techniques for recycling automotive shredder waste containing fibre-reinforced plastics is assessed. 3 refs.

VOEST ALPINE AG; AACHEN,UNIVERSITY

EUROPEAN COMMUNITY; GERMANY; WESTERN EUROPE

Accession no.512751

Item 346

Packaging Week

10, No.1, 5th May 1994, p.15

GERMANY WILL MEET PLASTICS TARGETS BY CHEMICAL RECYCLING

Hunt J

500,OOO tons of chemical recycling facility is due on stream in Germany during 1996, it is reported, making the packaging ordinance targets more of a reality, it is claimed. Around 250,000 tons capacity of mechanical recycling will also be available. This announcement follows the start up of BASF's pilot feedstock recycling plant at Ludwigshafen, the official opening of Veba Oel's facility in Bottrop, and RWE's claim that it will be using 70,000 tons of waste plastics to produce synthetic gas. Details of plant capacities are mentioned.

EUROPEAN COMMUNITY; GERMANY; WESTERN EUROPE

Accession no.512254

Item 347

Financial Times

No.32357, 4th May 1994, p.18

RUNNING ON RUBBISH

Brown-Humes C

Tests conducted in Finland and Sweden have indicated the viability of using waste paper and plastic packaging as a fuel in a conventional power plant rather than in a municipal solid waste incinerator. If the process is accepted, as much as 30 million tonnes of the 50 million tonnes of combustible packaging which Europe consumes each year could be used for power generation. The feasibility of the initiative is discussed, and its implications in terms of future power plant construction.

APME

EUROPE-GENERAL; EUROPEAN COMMUNITY; FINLAND; SCANDINAVIA; SWEDEN; UK; WESTERN EUROPE

Accession no.512233

Item 348

Communique

April 1994, p.3

REPORT ENDORSES PLASTIC INDUSTRY'S RECOVERY STRATEGY

A new study carried out by Dow Europe and published by APME indicates that both mechanical recycling and energy recovery have a vital role in ensuring the most environmentally efficient disposal for plastic bottles. The survey carried out by Dow uses the example of an HDPE bottle. Main conclusions of the report are indicated.

DOW EUROPE; APME

EUROPE-GENERAL

Accession no.512015

Item 349

Materials Reclamation Weekly

163, No.10, 7th May 1994, p.14

PLASTIC FIRMS PUSH FOR ENERGY RECOVERY

Research being carried out in Finland on the use of packaging derived fuels (PDF) is reported. Studies have assessed that refuse derived (RDF) and packaging derived fuels could save Europe 1 billion pounds sterling per year in non-renewable fossils fuels. Results are discussed of two research programmes which were commissioned to address the environmental impact of burning PDF with fossil fuels and the potential for PDF-based energy generation in Europe. Statistics are included.

ASSOCIATION OF PLASTICS MFRS.IN EUROPE
EUROPE-GENERAL; FINLAND; SCANDINAVIA; WESTERN EUROPE

Accession no.511967

Item 350
Materials Reclamation Weekly
163, No.10, 7th May 1994, p.4

BPF FIGHTS AMENDMENTS TO PACKAGING DIRECTIVE

The BPF is urging Euro MPs to vote against what it considers are undesirable amendments to the proposed Packaging and Packaging Waste Directive. Amongst the points of concern is the perceived hierarchy of recycling practices which places material recycling above energy recovery with no real consideration being given to the best recovery method for that particular material. Other proposed amendments include recovery targets, and mandatory use of recycled material content.

BPF
EUROPEAN COMMUNITY; UK; WESTERN EUROPE

Accession no.511964

Item 351
British Plastics and Rubber
Feb.1994, p.4-6

COUNTING THE COST OF PLASTICS RECYCLING

Appleboam V
Shell Chemicals Ltd.

This article discusses the options for the disposal of plastics waste and outlines the typical costs to society of various disposal methods such as mechanical recycling and energy recovery. On a highly selective basis, mechanical recycling of plastics can be economically viable, particularly during times of high virgin product prices. Once the waste is more widely distributed and becomes contaminated with other waste products, the costs involved increase considerably. Plastics have a very high intrinsic energy content that can be recovered in waste-to-energy plants, reducing demand for the oil, gas or coal that would otherwise be used for power generation.

EUROPEAN COMMUNITY; UK; WESTERN EUROPE

Accession no.511156

Item 352
Ends Report
No.230, March 1994, p.12-3

ICI LAUNCHES FEEDSTOCK RECOVERY PROGRAMME FOR ACRYLICS

It is reported that ICI Acrylics has launched a new recovery and recycling service for Perspex sheeting offcuts. Background details of the company and their recycling history are provided, and full details of the scheme are given.

ICI ACRYLICS; PEARCE SIGNS; ROHM & HAAS CO.; ATOHAAS
BELGIUM; EUROPE-GENERAL; EUROPEAN COMMUNITY; FRANCE; GERMANY; ITALY; PORTUGAL; UK; WESTERN EUROPE

Accession no.510183

Item 353
Composites Plastiques Renforces Fibres de Verre Textile
33, No.6, Nov/Dec.1993, p.21-4

THERMAL DECOMPOSITION OF FRP AND UTILISATION OF RESIDUE

Kitamura T; Hosokawa J; Kobayashi Y
Japan,Government Industrial Research Institute

Glass fibre-reinforced unsaturated polyester waste from boats was subjected to thermal decomposition in a water vapour atmosphere to obtain phthalic acid, styrene and glycols. Unsaturated polyester resins produced from the recovered phthalic acid showed reduced mechanical properties compared with a virgin sample, presumably due to the presence of benzoic acid formed by splitting of carbon dioxide from phthalic acid above 450C. The results suggested that recycling was possible by controlling the temperature below this level. The incineration ash was used to produce crystallised glass with a high flexural strength and relatively low bulk density, and which was suitable for use as a construction material. 5 refs.

JAPAN

Accession no.510114

Item 354
Plastics News(USA)
5, No.52, 28th Feb.1994, p.5

SPI WANTS COURT TO OVERTURN PYROLYSIS RULING

Gardner J

It is reported that the Society of the Plastics Industry wants an Oregon court to overturn a state Justice Department opinion that damages a chemical recycling project in the Pacific Northwest. The complaint seeks to protect the American Plastics Council's ability to comply with Oregon's Plastics container recycling law through pyrolysis of waste plastics at a Washington state plant. Details are given.

SPI; AMERICAN PLASTICS COUNCIL
USA

Accession no.510003

Item 355
Kautchuk und Gummi Kunststoffe
47, No.3, March 1994, p.202-6
German
ADVANCES IN CHEMICAL RECYCLING OF PLASTICS IN AUTOMOTIVE APPLICATIONS
Williams V
DuPont de Nemours International SA

Waste management options for plastics parts are analysed, using the disposal of automobiles as an example. The logistics of dismantling are outlined and the economic implications of mechanical recycling are compared with those of chemical recycling and energy recovery. The emphasis is on the chemical recycling of single polymer and mixed polymer waste streams, the activities of DuPont in both fields being highlighted.

SWITZERLAND; WESTERN EUROPE
Accession no.509601

Item 356
Kunststoffberater
38, No.4, April 1993, p.51-6
German
RECLAMATION OF USED PLASTICS BY HYDROGENATION
Holighaus R; Niemann K
Veba Oel AG

The Combi-Cracking process used at VEBA for the reclamation of used plastic materials by hydrogenation is seen as the most successful method currently available. The process is a development of the Bergius-Pier principle of high pressure hydrogenation in the liquid phase reactor used for liquefying coal and refinery waste products. The process is described using PE, PVC and polyamide as examples. 4 refs. (Lecture given at Seminar on Material Reclamation of Used and Waste Plastics at VDI, Wurzburg, January 1993).

EUROPEAN COMMUNITY; GERMANY; WESTERN EUROPE
Accession no.509381

Item 357
Plastics News(USA)
5, No.51, 21st Feb.1994, p.18
DOE PROCESS BASED ON HYDROLYSIS,PYROLYSIS
Ford T

The use is described of a process involving both hydrolysis and pyrolysis to recover caprolactam from nylon 6 used in carpet fibres. By means of precise temperature control and the use of a catalyst, nylon 6 can be isolated from the PP backing. The process has been developed by the National Renewable Resource Laboratory, and interest has been shown by AlliedSignal who are considering a cooperative research and development project.

US,NATIONAL RENEWABLE RESOURCE LABORATORY
USA
Accession no.509246

Item 358
Packaging Week
9, No.38, 7th April 1994, p.8
HAS THE FIZZ GONE OUT OF THE PET MARKET FOR METHANOLYSIS?

The slow take-up of methanolysis by soft drinks companies, including Coca-Cola, is discussed. Demand for depolymerised resins according to industry sources is put as low as 50 million pounds in 1993, with Eastman, Hoechst and Shell sharing some 150 million pounds of capacity. Coca-Cola, it is reported does not wish to become locked in to any particular technology, and like other industry experts, believes future developments will revolve around the three major techniques of depolymerisation, the use of multilayer, and the development of superclean flake technology which allows recycled PETP flake to be mixed with virgin material to give a bottle grade resin.

COCA-COLA CO.
USA
Accession no.509136

Item 359
Brussels, 1993, pp.12. 12ins. 15/12/93. 8(13)21
ROLE OF PLASTICS IN MUNICIPAL SOLID WASTE COMBUSTION. A TECHNICAL PAPER FROM A SERIES PRODUCED BY APME AND PWMI
Mark F E
Dow Chemical Europe; Association of Plastics Mfrs.in Europe; European Centre for Plastics in the Environment

The role of plastics in municipal solid waste combustion is discussed, and in particular, their fuel characteristics. Details are given of a state-of-art municipal solid waste combustor and the effects of plastics on the different combustion stages. In addition, the contribution of plastics to the total output spread over the various emission paths of solids, liquids and gases, is also assessed.

BELGIUM; EUROPEAN COMMUNITY; WESTERN EUROPE
Accession no.507878

Item 360
SPI Composites Institute's 48th Annual Conference. Conference Proceedings.
Cincinnati,Oh., 8th-11th Feb.1993, Paper 15-E. 627
RECYCLING THERMOSETS IN JAPAN UPDATE - TERTIARY REPORT
Kitamura T
Japan Reinforced Plastics Society
(SPI,Composites Institute)

The updated situation of Recycling Thermoset Composites in Japan is discussed. The dismantling or cutting up of ships, baths, water tanks, and automotive parts are covered. Recycling methods include pyrolysis and incineration. Details are also given of uses for recycled materials such as automotive parts, building materials and cement additives. 18 refs.

JAPAN

Accession no.507174

Item 361

European Chemical News

61, No.1606, 21st Feb.1994, p.23

BASF DELAYS DECISION ON FEEDSTOCK RECYCLING

BASF has delayed a decision on whether to press ahead with a DM300 million investment in a plastics feedstock recycling unit at Ludwigshafen. The decision to go ahead depends on the winning of a contract from DKR, a new recycling company for plastics packaging waste. The article supplies full details.

BASF CORP.; DKR

EUROPEAN COMMUNITY; GERMANY; WESTERN EUROPE

Accession no.505897

Item 362

Recycle '93. Conference Proceedings.

Davos, 22nd-26th March 1993, paper 23/4. 8(13)

RECYCLING: PRODUCT, FEEDSTOCK OR ENERGY? - A FUTURE VIEW

Dennison M T

Shell International Chemical Co.Ltd.

(Maack Business Services)

Waste management options for plastics are discussed in terms of an integrated approach which provides a balance between environmental benefit and overall cost to society. Plastics are safe in landfill, can be recycled into new plastics products or back into the chemical and oil feedstocks used in their original manufacture, or can be used as a source of energy to replace fossil fuels. The factors which must be considered when recycling plastics are discussed in detail. The energy recovery option will have to play a major role if a significant move away from landfill is to be achieved and it will need to be part of the integrated waste management and energy planning of each country in Europe.

EUROPEAN COMMUNITY; UK; WESTERN EUROPE; WESTERN EUROPE-GENERAL

Accession no.505826

Item 363

Recycle '93. Conference Proceedings.

Davos, 22nd-26th March 1993, paper 14/3. 8(13)

CHEMICAL RECYCLING OF USED PLASTIC MATERIALS: FIRST RESULTS

Gebauer M; Hofmann U

Leuna-Werke AG

(Maack Business Services)

Methods for recycling used plastic materials are reviewed. Emphasis is placed on the research projects into chemical recycling methods for used plastics at the Leuna location. These include development of a process for the thermal/thermooxidative pretreatment of used plastic materials, utilisation of pretreated used plastic materials in the visbreaker by gasification and by hydrogenation and the production of wax oxidates from pretreated used plastics. The results are discussed.

RWE AG; DARMSTADT,SCHOOL OF ENGINEERING; KOHLEOEL BOTTROP GMBH

EUROPEAN COMMUNITY; GERMANY; WESTERN EUROPE

Accession no.505788

Item 364

Recycle '93. Conference Proceedings.

Davos, 22nd-26th March 1993, paper 14/2. 8(13)

ADVANCES IN PRODUCING INDUSTRIAL GASES ORIGINATING FROM POST-CONSUMER PLASTIC WASTE

Lackner V; Michaeli W

Institut fuer Kunststoffverarbeitung

(Maack Business Services)

A reprocessing method for mixed plastics waste (preferably thermoplastic waste) is described in detail. This degradative extrusion process will allow the waste to be pretreated in a simple, rapid manner prior to chemical recycling. The results of dehydrochlorination tests, using model mixtures of mixed plastics during degradative extrusion, are discussed. Further investigations are being conducted to apply the results to real plastics mixtures and to scale up the process from the laboratory to an initial test plant.

PLASTICS WASTE MANAGEMENT INSTITUTE; KLOECKNER ER-WE-PA GMBH

EUROPEAN COMMUNITY; GERMANY; WESTERN EUROPE

Accession no.505787

Item 365

Recycle '93. Conference Proceedings.

Davos, 22nd-26th March 1993, paper 10/4. 8(13)

NEW DEVELOPMENTS IN THE CHEMICAL RECYCLING OF PLASTICS WASTE

Menges G; Brandrup J

Institut fuer Kunststoffverarbeitung

(Maack Business Services)

Different methods available for material recycling of plastics waste into raw materials for the chemical industry are reviewed and discussed. The technical problems, energy efficiencies and cost efficiencies of the processes are examined. 35 refs.

EUROPEAN COMMUNITY; GERMANY; WESTERN EUROPE

Accession no.505772

Item 366
Recycle '93. Conference Proceedings.
Davos, 22nd-26th March 1993, paper 6/5. 8(13)
LARGE SCALE ENERGY RECOVERY TRIALS ON POLYURETHANE, PET, ACRYLIC AND NYLON
Soederberg D J; Lenton R A; Boylett A R; Hicks D A
ICI Polyurethanes; ICI Films; ICI Engineering; ICI,Materials
(Maack Business Services)

The use of plastics as an energy source was demonstrated on a commercial scale at ICI Materials plastics manufacturing site in Dumfries, UK. This paper covers the preparation and use of pre- and post-consumer plastics as supplementary fuels in a circulating fluidised bed boiler specially designed for co-combustion with coal. Full emissions data on the 15% mixtures of individual plastics with coal are given, together with calculations of thermal efficiencies. Measurements by an independent body (British Coal Research Establishment) confirmed that the co-combustion of coal and plastic reduces some emissions compared with coal alone. Thermal efficiencies of around 80% were achieved and this heat was used effectively during the production of plastics. 7 refs.

BRITISH COAL RESEARCH ESTABLISHMENT
BELGIUM; EUROPEAN COMMUNITY; UK; WESTERN EUROPE

Accession no.505755

Item 367
Recycle '93. Conference Proceedings.
Davos, 22nd-26th March 1993, paper 6/4. 8(13)
COMBUSTION OF SHREDDED USED PACKAGING IN A MULTI-FUEL CFB BOILER
Frankenhaeuser M; Hiltunen M; Manninen H; Palonen J
Neste Oy Chemicals; Ahlstrom Corp.
(Maack Business Services)

This paper reports the results of an extensive co-combustion emissions testing programme designed for recovering energy from mixed plastics, refuse derived fuel and packaging derived fuel together with coal-containing primary fuel. The testing was done with a circulating fluidised bed boiler. 6 refs.

FINLAND; SCANDINAVIA; WESTERN EUROPE

Accession no.505754

Item 368
Recycle '93. Conference Proceedings.
Davos, 22nd-26th March 1993, paper 6/3. 8(13)
GASIFICATION: AN ECOLOGICAL RESPONSIBLE WAY FOR ENERGY GENERATION AND RECOVERY
Blessing J M
Blessing Business Services
(Maack Business Services)

Arcus' gasification combuster is described and the principles upon which it works are explained. This combuster combines solid fuel gasification with the burning of the lean gases produced on a small capacity scale. The types of solid fuels which can be used are listed and these include segregated municipal waste and industrial waste such as rigid PU foam and plastics mixed with other materials. Uses of the gas produced are included.

ARCUS-RECYCLING-SYSTEME GMBH; FH MINING
EUROPEAN COMMUNITY; GERMANY; WESTERN EUROPE

Accession no.505753

Item 369
Recycle '93. Conference Proceedings.
Davos, 22nd-26th March 1993, paper 5/4. 8(13)
MANAGEMENT OF SOLID WASTES IN PARIS AREA
Guillet R
Paris,Environment Protection Division

Figures are given for annual waste production in the Paris area and its composition is outlined. Many of the Paris area cities joined with Paris to create a solid wastes metropolitan authority for domestic waste treatment (SYCTOM). Three incineration plants burn 75% of the SYCTOM area solid wastes and the energy produced provides 43% of the energy consumed by the Paris urban heating network. Landfill is now expensive. There has been a reduction in the number of sites and French legislation prohibits landfill disposal of untreated solid wastes after 2002. A sorting unit at the landfill site was due to open in 1993 and another unit was planned for one of the incineration plants.

EUROPEAN COMMUNITY; FRANCE; WESTERN EUROPE

Accession no.505750

Item 370
Sheffield, c.1994, pp.2. 12ins. 1/3/94. 625-8(13)21
ENERGY RECYCLING OF PLASTIC FILMS
Packaging & Industrial Films Assn.

The recycling of plastic films is discussed with particular reference to energy recovery. Some statistics are included to demonstrate the size of energy recycling activities in Switzerland, Japan, Sweden and Denmark. Refuse derived fuels are also discussed, and the advantages of energy recycling are examined and compared to the processes involved in the recycling of film, and the costs involved in collection, sorting and cleaning.

EUROPEAN COMMUNITY; UK; WESTERN EUROPE

Accession no.505468

Item 371
Chemical Week
154, No.8, 2nd March 1994, p.20/2

PLASTICS RECYCLING: BACK TO FUELS AND FEEDSTOCKS
Rotman D; Chynoweth E

Chemical and thermal processes capable of recycling waste plastics back to fuels and petrochemical feedstocks are beginning to emerge as commercially viable alternatives to conventional recycling methods. Unlike established mechanical recycling, many of the thermal and chemical technologies can easily handle mixed plastic wastes and promise lucrative economic payoffs, yielding either high-quality oil or monomers with properties similar to virgin materials. Most observers agree that European producers are at the forefront in developing advanced recycling technology.

WORLD

Accession no.505089

Item 372
ECCM Recycling Concepts and Procedures.
Conference proceedings.
Bordeaux, 22nd-23rd Sept. 1993, p.59-64. 627-8(13)
LOW TEMPERATURE PYROLYSIS FOR CHEMICAL SEPARATION OF PLASTIC MIXTURES
Bockhorn H; Knumann R
Kaiserslautern,University
Edited by: Neitzel M; Lambert J C; Menges G; Kelly A
(European Association for Composite Materials; Commission of the European Communities)

The kinetics of thermal decomposition and depolymerisation of various polymers is discussed. The aim of the study was to find reaction conditions where different polymers can be separated from mixtures by decomposing them into their monomers or into pyrolysis products and where chlorine and/or nitrogen are eliminated from the polymers without forming toxic compounds. Data are given for PVC, PS, PE, and PP. 13 refs.

EUROPEAN COMMUNITY; GERMANY; WESTERN EUROPE

Accession no.504659

Item 373
ECCM Recycling Concepts and Procedures.
Conference proceedings.
Bordeaux, 22nd-23rd Sept. 1993, p.53-8. 627-8(13)
CHEMICAL RECYCLING OF PLASTICS
Menges G; Brandrup J
IKV
Edited by: Neitzel M; Lambert J C; Menges G; Kelly A
(European Association for Composite Materials; Commission of the European Communities)

Details are given of recycling plastic materials with emphasis on incineration. Its combination with chemical plants to use the flue gas as syngas is briefly mentioned. 5 refs.

EUROPEAN COMMUNITY; GERMANY; WESTERN EUROPE

Accession no.504658

Item 374
ECCM Recycling Concepts and Procedures.
Conference proceedings.
Bordeaux, 22nd-23rd Sept. 1993, p.41-6. 627-8(13)
RECOVERY OF MATERIAL AND ENERGY FROM THERMOSETTING PLASTICS
Pickering S; Benson M
Nottingham,University; British Rail Research
Edited by: Neitzel M; Lambert J C; Menges G; Kelly A
(European Association for Composite Materials; Commission of the European Communities)

Details are given of the development of energy and material recycling processes for thermosetting polymer composites. Applications in the cement industry and in coal fired fluidised bed combustion plants are discussed. 3 refs.

EUROPEAN COMMUNITY; UK; WESTERN EUROPE

Accession no.504656

Item 375
Brussels, 1993, pp.11. 12ins. 10/2/94. 6P-8(13)21
PACKAGING DERIVED FUEL (PDF) AS A SOURCE OF ENERGY. TECHNICAL PAPER
Frankenhaeuser M
Neste Chemicals
(Association of Plastics Mfrs.in Europe; European Centre for Plastics in the Environment)

Energy recovery from packaging waste is discussed, with particular reference to the co-combustion of mixed plastics with other conventional fuels such as wood, coal and peat. Experimental work is described in which a project was established to evaluate the possibility of energy recovery from a circulating fluidised bed boiler using packaging from different sources as fuel. The role of sulphur in the formation of PCDD/F in the combustion process was also studied.

BELGIUM; EUROPEAN COMMUNITY; WESTERN EUROPE

Accession no.503607

Item 376
Brussels, c.1994, pp.2. 12ins. 7/2/94. 8(13)21
WASTE TO ENERGY
European Centre for Plastics in the Environment

Energy recovery to reduce the amount of waste plastics going to landfill is shown by eco-balance studies to be more environmentally beneficial than recycling, it is reported. Advantages of the method and statistics to show current levels of activity are reported, and also the investment required by a company to operate such a process, and the running costs involved.

BELGIUM; EUROPEAN COMMUNITY; WESTERN EUROPE

Accession no.503356

Item 377

Brussels, c.1994, pp.2. 12ins. 7/2/94. 8(13)

FEEDSTOCK RECYCLING

European Centre for Plastics in the Environment

Current methods of feedstock recovery are reviewed. Brief details are given of pyrolysis, hydrogenation, gasification, and chemolysis. Activities of some European companies are briefly discussed in the areas of recycling mixed plastics waste and closed-loop recycling.

BELGIUM; EUROPEAN COMMUNITY; WESTERN EUROPE

Accession no.503355

Item 378

Plastics Recycling: Survival Tactics thru the 90's. Conference Proceedings.

Schaumburg, Il., 14th-16th June 1993, p.73-81. 8(13)

ADVANCED RECYCLING TECHNOLOGIES FOR PLASTICS

Meszaros M W

Amoco Chemical Co.

(SPE)

This paper discusses in depth advanced technologies for recycled materials from solid waste streams. Chemical depolymerisation, thermal depolymerisation, pyrolytic liquefaction, pyrolytic gasification, partial oxidation, and feedstock compatibility are all explained. The economic feasibility of the methods are considered.

BRITISH PETROLEUM; SHELL; HOECHST

EUROPEAN COMMUNITY; GERMANY; USA; WESTERN EUROPE

Accession no.502425

Item 379

Kunststoffe German Plastics

83, No.11, Nov.1993, p.21-3

SELECTIVE DISSOLUTION

Schurr U; Schneider M

The plastics portion in an average car, which in 1990 was still about 10 wt.%, is expected to increase to about 15 wt.% by the end of the century. In addition, the total number of cars is further expected to grow. It is therefore becoming increasingly necessary to recycle plastics waste from production as well as components from used cars. Legislation is aiming at increasing the reutilised portion to 20% by 1996. Insofar as the obtained used materials or parts consist of pure thermoplastics, their reutilisation creates little difficulty. There are also several promising approaches for reinforced glass fibre-reinforced products; aspects covered include adverse effects of composites on recycling, selective dissolution in a laboratory test, process technology proven in pilot plant, continuous operation improving the economic feasibility, a molecular sieve separating short-chain segments and selective dissolution not limited to SMA. 6 refs.

EUROPEAN COMMUNITY; GERMANY; WESTERN EUROPE

Accession no.501834

Item 380

Kunststoffe German Plastics

83, No.11, Nov.1993, p.18-20

FEEDSTOCK RECYCLING - A REVIEW

Vesper D E; Guhr U

It is claimed that only by fully considering all stages of the life cycle is it possible to arrive at a meaningful evaluation of a product. Thus the feasibilities and limitations of recycling are important, but are not the only aspects of the matter. Also discussed are the utilisation of plastics scrap as a raw material and energy carrier, and techniques under development. 11 refs.

EUROPEAN COMMUNITY; GERMANY; WESTERN EUROPE

Accession no.501833

Item 381

Plastics News(USA)

5, No.34, 18th Oct.1993, p.4

NRC STUDIES PYROLYSIS OF PLASTICS

Gardner J

Brief details are given of two proposed resolutions on the subject of pyrolysis of waste plastic. The first states that pyrolysis and other methods of chemically reprocessing post-consumer plastics is a suitable way of diverting waste from landfills. The second resolution, supported by environmentalists, states that pyrolysis only recovers plastic's energy value, and should not be viewed as recycling.

US,NATIONAL RECYCLING COALITION INC.

USA

Accession no.500602

Item 382

European Plastics News

20, No.11, Dec.1993, p.20

LCA POINTS TO ENERGY RECOVERY

The results of Dow Europe's life cycle analysis study into its HDPE bottle recycling scheme shows that mechanical recycling techniques can provide environmental benefits and underlines the value of incineration with energy recovery. The study shows that the system energy (the fuel energy value of the polymer plus the energy needed to manufacture and dispose of it) is virtually the same for both landfill and incineration disposal. If recycling is introduced, the total system energy is reduced by around 2.5% for each 10% of the bottles recovered from the MSW.

DOW EUROPE SA

SWITZERLAND; WESTERN EUROPE

Accession no.499442

Item 383
European Plastics News
20, No.10, Nov.1993, p.45-6
BURNING ISSUES
Chapelle A
APME

The plastics industry would like to see incineration with energy recovery more widely accepted as a way of dealing with municipal solid waste. Sweden recovers energy from 72% of its MSW to satisfy about 15% of its total district heating requirement and in Denmark, 60% of MSW is recovered for energy. The French Eco-Emballages system is fast becoming a policy model because it is based on a combination of solutions with realistic recycling targets balanced with energy recovery. The St Ouen energy recovery plant in Paris, which combusts 2,000 tonnes/day of waste produced in the surrounding areas, is given as an example.

WESTERN EUROPE; WESTERN EUROPE-GENERAL

Accession no.497966

Item 384
Recyclingplas VIII: Plastics Recycling as a Business Opportunity. Conference Proceedings.
McLean, Va., 3rd-4th June 1993, p.201-18. 8(13)
FROM PLASTICS WASTES BACK TO OIL
Holighaus R
Veba Oel AG; Kohleol-Anlage Bottrop
(Plastics Institute of America)

The conversion of plastics into clean liquid hydrocarbons includes cracking of the large polymer molecules as well as the separation of chlorine in case the waste material contains PVC. The liquid hydrocarbons generated are practically free of chlorine and can be directly reused as petrochemical feedstock. The use of the VCC process is described in detail.

EUROPEAN COMMUNITY; GERMANY; WESTERN EUROPE

Accession no.497937

Item 385
Recyclingplas VIII: Plastics Recycling as a Business Opportunity. Conference Proceedings.
McLean, Va., 3rd-4th June 1993, p.189-200. 8(13)
GASIFICATION PROCESS FOR MIXED PLASTICS WASTE
Simonsen K
Texaco Inc.
(Plastics Institute of America)

The Texaco Gasification process is a continuous, entrained flow, pressurised, non-catalytic partial oxidation process in which carbonaceous solids, liquids or gases react with oxygen. Gasification breaks the polymer chains and converts the hydrocarbons to their simplest forms. A detailed description is given of the process and its commercial application. The process is a commercially proven technology which has been in use for over 40 years. More than 100 commercial Texaco gasifiers have been licensed over this period. 18 refs.

USA

Accession no.497936

Item 386
Hoechst High Chem Magazine
No.13, 1993, p.37-9
CLOSED-LOOP RECYCLING INSTEAD OF DOWN-GRADING
Mueck K; Reuschel G
Hoechst AG

Hoechst has developed a chemical recycling plastic for Hostaform, a polyacetal engineering material. Post-use engineering parts and production scrap are recovered and converted back into the original monomers by depolymerisation. They are then repolymerised to form plastics with the same molecular structure as before, without loss of quality. The process at Hoechst's laboratory and pilot plant operations is outlined.

EUROPEAN COMMUNITY; GERMANY; WESTERN EUROPE

Accession no.497548

Item 387
European Plastics News
Nov.1993, German Plastics Supplement, p.21-2
LEARNING FROM GERMANY'S ENVIRONMENTAL DISASTER
Mader F
APME

The efficiency of the Duales System in Germany means that it will collect an estimated 400,000 tonnes of plastics packaging waste in 1993, far in excess of its legal obligation of 90,000 tonnes. This has over-stretched both the available recycling capacity and end-market demand. The more flexible approach being taken by France and Italy looks more likely to provide the answer to waste management problems. French policy is based on an integrated strategy in which re-use, recycling and energy recovery combine to optimise the use of resources.

EUROPEAN COMMUNITY; GERMANY; WESTERN EUROPE

Accession no.497215

Item 388
Reuse/Recycle
23, No.8, Aug.1993, p.60
NOVEL PROCESS CONVERTS PLASTICS INTO OIL

Brief details are given of an oil recovery process currently being researched at the Institute for Mining and Materials Research, Kentucky University. The research project is part of a larger programme carried out by a 5 university, Dept. of Energy funded project. In this process oil is produced from plastic liquefaction, details of which are given.

KENTUCKY,UNIVERSITY
USA

Accession no.495821

Item 389
Chemical and Engineering News
71, No.40, 4th Oct.1993, p.11-4
ADVANCES IN FEEDSTOCK RECYCLING OFFER HELP WITH PLASTIC WASTE
Layman P L

Details are given of processes currently under development in Europe and the US for recycling plastics waste into feedstocks. In particular, the efforts of BP Chemicals and Shell are reported who have plans to form consortia of chemical companies. Problems associated with costs and logistics are discussed, which are preventing full-scale commercialisation.

BP CHEMICALS LTD.; SHELL CHEMICAL CO.
EUROPE-GENERAL; EUROPEAN COMMUNITY; UK; USA; WESTERN EUROPE

Accession no.495039

Item 390
Packaging Waste Management: Day 1. Conference Proceedings.
Gatwick, 19th-20th Nov.1992, Paper 5. 8(13)
PLASTIC WASTE - A SOURCE OF USEFUL ENERGY
Behrendt P F
Dow Chemical Co.
(Pira International)

In the UK every year, 30 million tonnes of municipal solid waste is collected. 7% by weight of this is plastics - which contributes about a quarter of the total energy which can be recovered. This paper looks at the European Commission's integrated concept for plastic disposal - source reduction, material recycling, chemical recycling, energy recovery, incineration and landfill. Particular emphasis is given to energy recovery. The current situation in the UK is compared with that of Western Europe.

OTTO GMBH; AUTOBAR FEDERATION; SHELL; SAVE-A-CUP RECYCLING CO.; BP CHEMICALS LTD.; BRITISH PLASTICS FEDERATION; NEWCASTLE,CITY COUNCIL; ENERGY FROM WASTE PRODUCERS; EUROPEAN COMMISSION
DENMARK; EUROPEAN COMMUNITY; FRANCE; GERMANY; NETHERLANDS; SCANDINAVIA; SWEDEN; UK; WESTERN EUROPE; WESTERN EUROPE-GENERAL

Accession no.493814

Item 391
Chemistry & Industry
20, No.8, Sept.1993, p.699
TURNING PLASTICS INTO OILS
Burke M

Researchers at the University of Kentucky estimate that, at the current rate of plastics waste disposal in the USA, over 80 million barrels of oil could be produced. Unfortunately, only 3.7% of plastics waste was recycled out of 22 million tons collected in 1990; the rest was incinerated or ended up in landfills. Details are given.

KENTUCKY,UNIVERSITY
USA

Accession no.493065

Item 392
Ends Report
No.221, June 1993, p.32-3
GOVERNMENT TO USE RCEP REPORT IN A NEW PUSH FOR INCINERATION

The Government is reported as being poised to invoke the favourable assessment of incineration by the Royal Commission on Environmental Pollution, to give the technique support as part of its recycling strategy for municipal waste. An energy recovery target is still under consideration. Brief details are given.

UK,ROYAL COMMISSION ON ENVIRONMENTAL POLLUTION
EUROPEAN COMMUNITY; UK; WESTERN EUROPE

Accession no.492807

Item 393
Recycle '92 Conference Proceedings.
Davos, 7th-10th April 1992, p.3/4-1 - 3/4-18. 8(13)
CO-COMBUSTION OF MIXED PLASTICS WITH COAL IN A BUBBLING FLUIDISED BED BOILER
Frankenhaeuser M
Neste OY
(Maack Business Services)

Mixed plastics waste appears to be well suited for use in energy recovery, either as a co-combustion fuel in a power plant designed for solid fuels, or as the sole fuel in specially designed plants. This paper reports test results on the co-combustion of mixed household plastics with coal. The tests were performed in a bubbling fluidised bed low-pressure steam boiler. The results show that both inorganic and organic total specific emissions were lower for mixed household plastics than for coal. Tabulated data are presented. 3 refs.

FINLAND; SCANDINAVIA; WESTERN EUROPE

Accession no.492604

Item 394
Recycle '92 Conference Proceedings.
Davos, 7th-10th April 1992, p.3/3-1 - 3/3-6. 8(13)
HURDLES IN THE THERMAL CRACKING OF PLASTIC WASTE

Huybrechts S
BP Chemicals Ltd.
(Maack Business Services)

In examining the technical options for plastic waste management, chemical recycling appears to be the least developed and most difficult. In this paper, BP Chemicals sets out its analysis of the factors that will determine the choice of chemical recycling process technology. From this a process concept based on thermal cracking is developed and the hurdles to be overcome before such a process can be realised is discussed.

EUROPEAN COMMUNITY; UK; WESTERN EUROPE

Accession no.492603

Item 395

Recycle '92 Conference Proceedings.
Davos, 7th-10th April 1992, p.3/2-1 - 3/2-12. 8(13)

TERTIARY RECYCLING OF POLYOLEFINS

Meszaros M
Amoco Chemical Co.
(Maack Business Services)

Recycling plastics back to hydrocarbons (tertiary recycling) may offer significant promise. Three different options currently being evaluated for the tertiary recycling of polyolefins - refinery recycling, pyrolysis and depolymerisation - are discussed.

USA

Accession no.492602

Item 396

Recycle '92 Conference Proceedings.
Davos, 7th-10th April 1992, p.3/1-1 - 3/1-15. 8(13)

FEEDSTOCK & ENERGY RECOVERY

Stannard D C
Shell International Chemie BV
(Maack Business Services)

Material recycling is the objective for every material, but at some point reuse or collection, separation and further recycling will no longer yield a useful product. The so-called "plastic waste" still contains a high calorific value which can be recovered to produce heat or electricity. Even better it may be possible to recover the chemical feedstock originally manufactured from oil. These two possibilities are reviewed.

EUROPEAN COMMUNITY; NETHERLANDS; WESTERN EUROPE

Accession no.492601

Item 397

Ends Report

No.216, Jan.1993, p.13-4

BP SEEKS PARTNERS TO TURN WASTE PLASTICS BACK TO OIL

BP Chemicals' plans for a plastics recycling consortium are detailed. The company, along with 4 other undisclosed European petrochemical companies, plans a commercial scale unit by the end of 1997 which involves polymer cracking, a process by which polymer feedstocks are recycled.

BP CHEMICALS LTD.

EUROPE-GENERAL; EUROPEAN COMMUNITY; UK; WESTERN EUROPE

Accession no.492491

Item 398

Food, Cosmetics & Drug Packaging

Sept.1993, p.8

INCINERATION IS BEST DISPOSAL FOR SNACK FOOD PACKAGING

Warwick Research Institute

Key findings are reported from a study by David Brown of Warwick Research Institute into the environmental aspects of plastics packaging used for snack foods. The four options of reduction, re-use, recycling and recovery are discussed. The original research examines OPP as the most commonly used plastic for snack food packaging.

BODY SHOP

EUROPEAN COMMUNITY; UK; WESTERN EUROPE

Accession no.492482

Item 399

Plastics World

51, No.9, Sept.1993, p.14

STEEL-MAKING PROCESS CONSUMES PLASTICS SCRAP

Miller B

Voest-Alpine Industrieanlagenbau GmbH has developed, and has plans to commercialise, a process for metallurgical recycling of scrap steel that is claimed to be an economically feasible and environmentally sound solution to the problem of disposing of plastics from automotive shredder residue. The process eliminates the need for separation of steel and plastic, by destroying the intermixed plastic as it is fed to the melting furnace. Details are given of the preheating conditions which are controlled to safely decompose the plastic and from which process recovered energy helps to reduce overall fuel costs of the process.

VOEST-ALPINE INDUSTRIEANLAGENBAU GMBH

AUSTRIA; WESTERN EUROPE

Accession no.492380

Item 400

Plastics and Rubber Asia

8, No.47, Aug.1993, p.20

PLASTICS INTO FUEL OIL

The Japan Small Business Corp. has developed a waste plastics reprocessing system to convert plastics in domestic waste into fuel oil. The system uses an oil refining catalyst in addition to pyrolysis and enables fuel oil to be recovered at a high yield. The article supplies details of the reprocessing system which is expected to contribute to promoting the recycling of waste plastics.

NIPPON STEEL CORP.; JAPAN SMALL BUSINESS CORP.; FUJI INDUSTRIAL RECYCLE CORP.; SHINAGAWA FUEL CO.LTD.
JAPAN

Accession no.491866

Item 401
Plastics and Rubber Weekly
No.1499, 21st Aug.1993, p.1
FACE THE FACTS: BURN UK'S DOMESTIC WASTE

With reference to the latest UK government's attempts to tackle the problem of the disposal and/or recycling of packaging waste, comments are reported from Cameron McLatchie, chief executive of British Polythene Industries. He calls for a landfill levy and increased use of incineration with energy recovery. According to recent studies, the capacity for mechanical recycling is presently insufficient. Statistics relating to waste production are included and the case for incineration in the UK is propounded.

BRITISH POLYTHENE INDUSTRIES
EUROPEAN COMMUNITY; UK; WESTERN EUROPE

Accession no.490232

Item 402
Chimica e l'industria
75, No.3, March 1993, p.227-8
Italian
PYROLYSIS FOR THE DISPOSAL OF PLASTICS WASTE

Details are given of a pyrolysis process developed by BP Chemicals for the recovery of raw materials from plastics waste.

BP CHEMICALS LTD.
EUROPEAN COMMUNITY; UK; WESTERN EUROPE

Accession no.489716

Item 403
Plastics News(USA)
5, No.16, 14th June 1993, p.4
DEPOLYMERISATION PROPONENTS OPTIMISTIC
Gardner J

Proponents of a technology that converts plastics into chemical feedstocks for use as new resin or fuel, are optimistic that it will eventually be accepted as recycling. The technologies and targets vary, but, according to representatives, the conversion of plastics into feedstocks is a feasible recycling technology for wastes that do not presently have markets through mechanical recycling. The article supplies details of the depolymerisation technologies, their advantages and disadvantages.

PLASTICS INSTITUTE OF AMERICA; FAGAN TECHNOLOGY INC.; FUJI RECYCLE INDUSTRY KK; TEXACO INC.
USA

Accession no.489454

Item 404
Kunststoffe Plast Europe
No.2, June 1993, p.166/70
French; English
RECYCLING OF SMALL PARTS
Prautsch G

The recycling of small automotive plastic components is discussed where separation of the plastics material from metal is involved. High grade engineering plastics, in particular nylon 6 and 66, are being chemically recycled by Du Pont within the framework of its global programme for improving the environmental compatibility of polymers. Mention is made of the patented technology which uses nylon moulded parts and fibres, although no details are given.

DU PONT DE NEMOURS E.I.,& CO.INC.
EUROPEAN COMMUNITY; GERMANY; WESTERN EUROPE

Accession no.487053

Item 405
Materials Reclamation Weekly
161, No.17, 12th June 1993, p.13/15
CHEMICAL TECHNOLOGY TO RECOVER QUALITY FEEDSTOCKS FROM USED PLASTICS

Information is presented in some detail on Du Pont Polymers' process for recovery of used polyesters from products such as automotive bumpers. The process is reported to use a methanolysis technique and is reported to be able to separate pure monomer ingredients from reinforced, painted or otherwise contaminated polymers.

DU PONT DE NEMOURS E.I.,& CO.INC.,POLYM.PROD.DEPT.
USA

Accession no.483392

Item 406
Recyclingplas VII: Plastics Recycling as a Business Opportunity. Conference Proceedings.
Arlington, Va., 20th-21st May 1992, p.206-20. 8(13)
VACUUM PYROLYSIS OF AUTOMOBILE SHREDDER RESIDUE
Roy C; Dubuc M
Pyrovac Institute Inc.; ETP Technologies Inc.
(Plastics Institute of America)

It is reported that between 10 and 11 million cars are taken apart every year in North America; once the metallic portion has been recovered from the crushing operation, the remainder - representing over 20% of the total mass or 3.5 million tons - must be eliminated. Pyrovac, in collaboration with Universite Laval and under the sponsorship of E.T.P. Technologies, recently undertook a research programme to recover valuable products from this Automobile Shredder Residue. A process flow sheet for a 4000 kg/hr plant is presented, with fixed capital investment estimated to be 5 million US dollars. Details are given.

USA

Accession no.480788

Item 407

Recyclingplas VII: Plastics Recycling as a Business Opportunity. Conference Proceedings.
Arlington, Va., 20th-21st May 1992, p.175-90. 8(13)

INNOVATIVE PYROLYTIC APPROACHES TO THE RECYCLING OF PLASTICS TO MONOMERS

Evans R; Tatsumoto K; Czernik S; Chum H L
US,National Renewable Energy Laboratory
(Plastics Institute of America)

An attempt is made to identify conditions for the production of monomers and high-value chemicals from mixtures of waste plastics, by the use of controlled thermal processes. Feedstock presorting and product purification are minimised by controlling reaction conditions. Target waste streams are from plastics manufacturing, consumer product manufacturing and post-consumer sources. The most promising applications to date are the recovery of caprolactam from waste in the manufacture of nylon 6 carpet, diamine derivatives from waste PUs and dimethyl terephthalate from mixed wastes containing PETP. 23 refs.

USA

Accession no.480786

Subject Index

D

E

H

I

J

K

L

M

N

O

P

Q

R

S

T

U

V

W

X

Y

Z

DOCUMENTS DIRECT

(Document Delivery Service)

The Polymer Library (www.polymerlibrary.com) is the world's most comprehensive collection of information on the rubber, plastics, composites and adhesives industries. The fully searchable database covers approximately 500 regular journals as well as conference proceedings, reports, books, company brochures and data sheets.

Almost all the articles selected for the database can be ordered in full text through our document delivery department. Non-patent requests are usually despatched within 24 hours of receipt (Monday to Friday).

- We have a large collection of literature directly related to the industries we serve and can offer a personal service with minimal bureaucracy, based on detailed knowledge of our stock.
- Many of the documents held at Rapra are not available via other services. This is particularly the case for our extensive and unique collection of company literature and data sheets.
- We offer a *fast turnaround service* (within one working day) combined with a range of delivery options. Some full text documents are available as PDF files which can be downloaded immediately

SPEED OF DELIVERY

Non-patent documents are despatched from Rapra within 24 hours of receipt (Monday - Friday) of request using first class mail within the UK, and airmail for the rest of the world. If you request e-mail or fax service, delivery will be within hours anywhere in the world.

HOW TO ORDER

Orders can be made by post, fax, telephone, e-mail, on-line via the website database (http://www.polymerlibrary.com), or through an online host.

When ordering please include your full company details and which documents you require, quoting one of the following:

1. Accession Number or Copyquest number or,
2. Full Bibliographic Details

Please include which payment method you wish to use and how you wish to receive the article (i.e. e-mail, post, fax, etc.)

Documents can be ordered from Rapra online using the appropriate command of your online host. In this case we will issue you with an invoice and statement every three months.

For further information, please see www.rapra.net/absdocs/copyquest.htm or contact Sheila Cheese or Jackie McCarthy on +44 (0)1939 250383 or e-mail documents@rapra.net.

PLEASE TURN OVER FOR PAYMENT METHOD OPTIONS AND ORDER FORM

CREDIT CARD PAYMENTS

This is preferable for people who only intend to use our service occasionally. The prices are per copy inclusive of postage and packaging if appropriate. Pre-payment is required by Credit Card payment.

Delivery Method	Price
E-mail, Ariel, Fax, First Class / Airmail Post	£10 / Approx. US$15 or €17 (UK & Overseas) + VAT
PDF files	Prices – Please refer to online order form for details (these may vary – each publisher sets price)

ANNUAL DEPOSIT ACCOUNTS

A more cost effective way to use our service is to open an annual deposit account. (Web subscribers can use their web subscription for both web access and document delivery)

The minimum amount required to open an account is £200 / Approx. US$290 or €330 (UK & Overseas)

Delivery Method	Price
E-mail, Ariel, Fax, First Class / Airmail Post	Documents are only priced at £7 / Approx. US$10 or €12 per item
PDF files	Prices – Please refer to online order form for details (these may vary – each publisher sets price)

Please Note: Any money remaining in an annual deposit account after 12 months is void.

Patents are charged at a standard price of £10 / Approx. US$15 or €17.

ORDER FORM

❑ I would like to open/renew a deposit account for the following amount ______________________

❑ I would like to order the following documents ______________________

____

PAYMENT

Name: ______________________

Company: ______________________

Address: ______________________

____

____

Postcode: ____________ Country: ____________

Telephone: ____________ Fax: ____________

IMPORTANT - Value Added Tax (VAT)

The above prices do not include VAT. Customers in EU member countries may be liable to pay VAT if their Registration Number is not supplied. Please enter your EU Registration Number (VAT - BTW - IVA - TVA - MWST - MOMS - FPA) below:

VAT Number:

❑ **Remittance enclosed**
(use only for opening or renewing annual deposit accounts)

(If paying by cheque, please make payable to Rapra Technology Ltd. in £ Sterling/US$/Euros via UK banks only or make payment direct to Account No: 05625025, Sortcode: 55-50-05, National Westminster Bank Plc, 8 Mardol Head, Shrewsbury, Shropshire, SY1 1HE, UK)

❑ **Please invoice my company** *(use only for opening or renewing annual deposit accounts)*

❑ **Please deduct from my annual deposit account** *(use this option when ordering documents if you already have a deposit account)*

Account Number ______________________

❑ **Please charge my credit card**

American Express / Visa / Mastercard *(delete as appropriate)*

Card Number:

Signature: ______________ Exp. date: ____________

Rapra Technology Limited
Shawbury, Shrewsbury, Shropshire SY4 4NR, United Kingdom
Tel. +44 (0)1939 250383 Fax: +44 (0)1939 251118
E-mail: documents@rapra.net